Radio Modernisms

This collection interrogates and stimulates deep, cross-disciplinary engagement with the various understandings and interplays of 'radio modernisms' from the early decades of the twentieth century through to the 1950s.

Academics from a range of disciplines explore their common interests in the richness and heterogeneity of BBC Radio's imaginative programming – in terms of sound; as cultural events from specific moments in time; as team creations; as something experienced live in the domestic context; and as cultural works that, in many cases, attracted a certain canonical pedigree. Radio modernisms are, as these chapters demonstrate, a combination of the particular, the contingent, and the contextual. More than a decade after the publication of the first scholarly works to yoke together 'modernism' and 'radio', this collection emphasises the plurality of 'modernisms' as a defining aspect of contemporary BBC historiography. The authors bring multiple lenses to bear – including race, gender, and transnationalism – in order to (re)locate twentieth-century radio programming in broad, expansive contexts. They also underline the dynamic entanglements of radio – and radiogenic feature programmes, in particular – with other kinds of media and cultural forms and formats, reframing radio as a site of and vehicle for remediation and intermediality.

In examining the myriad ways in which radio gave shape to new modernities, and both evolved and constituted new forms of modernism, this collection offers fresh perspectives on the interconnected significance of 'radio modernisms' within the socio-cultural, literary, and political landscapes of twentieth-century Britain. This book was originally published as a special issue of *Media History*.

Aasiya Lodhi is a Senior Lecturer in the School of Media and Communication at the University of Westminster, UK. She was previously an arts and current affairs producer at BBC Radio 3, BBC Radio 4, and BBC World Service.

Amanda Wrigley is a Postdoctoral Researcher in the Department of Film, Theatre and Television at the University of Reading, UK. She is also a Visiting Fellow in the School of Arts and Cultures at the Open University, UK, and a Fellow of the Royal Historical Society.

Radio Modernisms

Features, Cultures and the BBC

Edited by
Aasiya Lodhi and Amanda Wrigley

Routledge
Taylor & Francis Group
LONDON AND NEW YORK

First published 2020
by Routledge
2 Park Square, Milton Park, Abingdon, Oxon, OX14 4RN

and by Routledge
52 Vanderbilt Avenue, New York, NY 10017

Routledge is an imprint of the Taylor & Francis Group, an informa business

British Library Cataloguing in Publication Data
A catalogue record for this book is available from the British Library

ISBN13: 978-0-367-36765-7

Typeset in Myriad Pro
by Newgen Publishing UK

Publisher's Note
The publisher accepts responsibility for any inconsistencies that may have arisen during the conversion of this book from journal articles to book chapters, namely the inclusion of journal terminology.

Disclaimer
Every effort has been made to contact copyright holders for their permission to reprint material in this book. The publishers would be grateful to hear from any copyright holder who is not here acknowledged and will undertake to rectify any errors or omissions in future editions of this book.

Contents

Citation Information

The chapters in this book were originally published in *Media History*, volume 24, issue 2 (May 2018). When citing this material, please use the original page numbering for each article, as follows:

Chapter 1
Introduction: Radio Modernisms: Features, Cultures and the BBC
Aasiya Lodhi and Amanda Wrigley
Media History, volume 24, issue 2 (May 2018), pp. 159–165

Chapter 2
Radio's Vernacular Modernism: The Schedule as Modernist Text
Kate Lacey
Media History, volume 24, issue 2 (May 2018), pp. 166–179

Chapter 3
Waves: Aestheticism, Radio Drama and Virginia Woolf
Todd Avery
Media History, volume 24, issue 2 (May 2018), pp. 180–193

Chapter 4
BBC Features, Radio Voices and the Propaganda of War 1939–1941
Alex Goody
Media History, volume 24, issue 2 (May 2018), pp. 194–211

Chapter 5
Making Waves: Una Marson's Poetic Voice at the BBC
Leonie Thomas
Media History, volume 24, issue 2 (May 2018), pp. 212–225

Chapter 6
'Countries in the Air': Travel and Geomodernism in Louis MacNeice's BBC Features
Aasiya Lodhi
Media History, volume 24, issue 2 (May 2018), pp. 226–238

Chapter 7

Who's Listening to Modernism? BBC Features and Audience Response
Alexandra Lawrie
Media History, volume 24, issue 2 (May 2018), pp. 239–251

Chapter 8

Intermedial Relationships of Radio Features with Denis Mitchell's and Philip Donnellan's Early Television Documentaries
E. Charlotte Stevens and John Wyver
Media History, volume 24, issue 2 (May 2018), pp. 252–265

Chapter 9

Afterlives of BBC Radio Features
Amanda Wrigley
Media History, volume 24, issue 2 (May 2018), pp. 266–282

Chapter 10

Afterword: Radio Modernisms: Features, Cultures and the BBC
David Hendy
Media History, volume 24, issue 2 (May 2018), pp. 283–287

Notes on Contributors

Todd Avery is a Professor of English at UMass Lowell, US. He writes extensively on British modernism, with a secondary emphasis in cultural studies and a perennial concern with relations among literature, ethical philosophy, and emergent technologies.

Alex Goody is a Professor of Twentieth-Century Literature and Culture at Oxford Brookes University, UK. She specialises in the areas of modernist and avant-garde studies, women's writing, technology and literature, and media and cultural studies.

David Hendy is a Professor of Media and Cultural History at the University of Sussex, UK. He has written extensively on the role of sound, images, and communication in human culture, and on the connections between modern 'mass' media and popular life and thought in the nineteenth-, twentieth-, and twenty-first centuries.

Kate Lacey is a Professor of Media History and Theory at the University of Sussex, UK. Her research interests include twentieth-century German history, broadcasting history, and gender studies.

Alexandra Lawrie is a Senior Lecturer at the University of Edinburgh, UK. Her research interests are in contemporary British and American fiction, and she has written on the rise of English in extramural institutions.

Aasiya Lodhi is a Senior Lecturer in the School of Media and Communication at the University of Westminster, UK. Her research interests include post-war and post-empire British radio, and radiogenic forms and their history, especially features.

E. Charlotte Stevens is a Research Fellow at Birmingham City University, UK. Her research interests include television history, archival practice, and amateur historiography.

Leonie Thomas is completing her PhD at the University of Bristol, UK. She is interested in radio studies, female voices at the BBC, and mid-century British writers.

Amanda Wrigley is a Postdoctoral Researcher in the Department of Film, Theatre and Television at the University of Reading, UK. She is a cultural historian who has written extensively on literary, dramatic, and historical themes in twentieth-century British radio and television programming.

John Wyver is a Professor of the Arts on Screen at the University of Westminster, UK. He is a historian of television with special interests in broadcasting, film, and the arts.

INTRODUCTION: RADIO MODERNISMS
Features, Cultures and the BBC

Aasiya Lodhi and **Amanda Wrigley**

This special issue arises from a one-day conference on the topic of 'Radio Modernisms: Features, Cultures and the BBC' that we held on 19 May 2016 at the British Library, with the support of the Communications and Media Research Institute at the University of Westminster. Almost all contributors to this issue gave papers at the conference; others who spoke or who had planned to speak (Hugh Chignell, Henry Mead, Kate Murphy and Paul Wilson) enriched the discussions in welcome ways, and we remain grateful for their contributions. The talk by Paul Wilson, Curator of Radio at the British Library, brought the significant issue of archives—and their preservation, curation and accessibility—centre stage; his championing of the series of public listening events ('Louis MacNeice: Radio Writer and Producer', curated by Amanda Wrigley) across May-June 2016 meant that the conference's papers and discussions were followed by an act of communal listening to, and discussion of, a little-known example of MacNeicean radio.

The conference was attended by around fifty participants from richly diverse areas of expertise including practice-based research and the historical study of architecture, broadcasting, classics, drama, imperialism, literature, modernism, musicology, radio, sound, television and transnationalism. The enthusiastic engagement of participants across the day underscored the sense that 'radio modernisms' as an idea represented a productive meeting-point for the exploration of common questions from a broad range of perspectives. This not only indicated a hot interdisciplinary topic but it also mirrored distinctive aspects of the conference's particular focus—the programmes, aesthetics, personnel and creative practices of the BBC Features Department in the middle stretch of the twentieth century.

In truth, we did not anticipate that this topic would resonate so widely. The idea for the conference had arisen from a specific desire to engage two scholarly constituencies more closely in dialogue with each other. We had observed with excitement the recent flowering of interest from scholars of English literature in 'literary radio' (for want of definitive terminology). The radio programmes of interest to these scholars either adapt and realise in sound modernist works in print, bring to light the radio writings of canonical and more marginal modernist authors, or are radiogenic creations (often feature programmes) that exploit radio technology in a way that engages with the concerns and aesthetics of literary modernism and modernity itself. But the focus is often concentrated on the text and the writer. The second constituency comprises broadcasting historians in the UK—especially those, like ourselves, who have literary backgrounds and are actively researching the literary cultures and modernist aesthetics of BBC radio in the middle decades of the twentieth century. This constituency is rather diffusely spread across a variety of institutions, research centres and departments (of, for example, communications, cultural studies, education, English literature, modern history, media history, media studies, television and theatre). On the one hand, then, the practical aim of the conference was to enhance the sense of connection amongst those based in the UK with research interests in BBC Radio's imaginative programming from the mid-twentieth century; and, on the other, the intellectual project was to explore how we may best share and learn from each others' methodologies and sets of expertise with the aim of strengthening scholarship across disciplinary boundaries.

The conference was held at the ten-year point since the 2006 publication of Todd Avery's landmark book *Radio Modernism: Literature, Ethics, and the BBC, 1922-1938* (and we were delighted that he was able to give the keynote address). A number of book-length studies by literary modernist scholars followed: for example, many of the essays in Debra Rae Cohen, Michael Coyle and Jane Lewty's 2009 collection *Broadcasting Modernism* and Matthew Feldman, Erik Tonning and Henry Mead's 2014 *Broadcasting in the Modernist Era* are concerned with modernist *auteur* engagements with radio, as are Melissa Dinsman's 2015 *Modernism at the Microphone*, Emily C. Bloom's 2016 *The Wireless Past* and David Addyman, Matthew Feldman and Erik Tonning's *Samuel Beckett and BBC Radio* (2017, eds). Without neglecting their fundamentally important skills of literary criticism and textual analysis, literary scholars are increasingly seeking to anchor their work more broadly. Bloomsbury's Historicizing Modernism series (in which Feldman, Tonning and Mead's 2014 volume and Dinsman's 2015 monograph are published) emphasises the importance of historical sources beyond the modernist text itself. Another recent collection, Debra Rae Cohen and Michael Coyle's special issue of *Modernist Cultures*, titled 'Broadcast Traces / Tracing Broadcasting: Modernism and Radio' (2015), perceptively highlights the need for modernists interested in literary radio to 'move beyond the simple limning of relations between single authors and radio' as they introduce new work that is intended to '[trace] plural histories, multiple conversations, and feedback loops' (2). The latest monograph on the shelf of literary radio tomes, published almost simultaneously with this special issue, is Ian Whittington's 2018 *Writing the Radio War: Literature, Politics and the BBC, 1939-1945*. Whittington conforms to the customary one author per chapter arrangement, but seeks to represent the 'broad literary coalition' of wartime radio writers and in doing so navigates cross-currents of broader cultures and politics assuredly (4). *Writing the Radio War* is about literary radio but its foundations are much broader.

This flourishing interest in literary radio revives aspects of a rich seam of scholarship from the 1980s that had been foundational for our own research. For example, *British Radio Drama*, edited by John Drakakis in 1981, offered essays by Christopher Holme on Louis MacNeice, Peter Lewis on Dylan Thomas, Donald A. Low on Susan Hill and Dorothy L. Sayers, Frances Gray on Giles Cooper, Roger Savage on Henry Reed, Katharine Worth on Samuel Beckett and David Wade on radio drama from the 1960s. *Radio Drama*, edited by Peter Lewis, focused less on individual *auteurs*, with contributors (English literary scholars mainly, with some practitioners) bringing a more generic, theoretical or practice-based lens to the topic of drama, fiction and features. In 1982, Ian Rodger's *Radio Drama* brought his practice-based and critical expertise to an interpretation of the development of the form across three decades from the early 1920s. Kate Whitehead's 1989 monograph, *The Third Programme: A Literary History*, offered a tightly integrated discussion of the institutional contexts for the work of poets, playwrights and novelists for the BBC's Third Programme network, 1946–1970. The qualities that these works from the 1980s have in common include a serious engagement with creative processes within institutional frameworks, the aesthetics and intermediality of literary radio and an assessment with the evidence for programmes and programme-making preserved in the BBC Written Archives Centre. They offer, in short, valuable historiographical models which were soon underpinned by excellent cultural and social histories of pre-war broadcasting: D. L. LeMahieu's 1988 *A Culture for Democracy* and Paddy Scannell and David Cardiff's 1991 *A Social History of British Broadcasting*.

Those currently engaged in historical research on literary radio who style themselves as historians of broadcasting, culture, media, society, etc, do engage in close analysis of programmes (including the text, whether existing as audio record or script) but they are keen to decentre that text through the application of historicist methodologies to the investigation of its contexts, such as collaborative practices amongst producers and other creative personnel involved in realising scripts through sound; to investigate the institutional frameworks and policies informing programme choices; and to assess the properties and grammar of the soundscape itself which (when extant and available for researchers to access) conveys a fuller sense of the original audience's sensory experience of a programme than any script alone can offer. In broadcasting history more generally, there has been attention paid to listenerships (Kate Lacey's 2013 *Listening Publics*; Amanda Wrigley's 2015 *Greece on Air*), national identity (Thomas Hajkowski, *The BBC and National Identity in Britain*, 2010), transnationalism (Michele Hilmes, *Network Nations*, 2011), the soft diplomacy and cold war politics of overseas broadcasting (Simon Potter, *Broadcasting Empire*, 2012; Alban Webb, *London Calling*, 2014), the role of women at the BBC (Kate Murphy, *Behind the Wireless*, 2016) and architecture and the built environment to 1945 (Shundana Yusaf, *Broadcasting Buildings*, 2014), to name just a few recent perspectives that illuminate the rich heterogeneity of radio's histories and deepen our knowledge of broadcasting, and the BBC in particular. Current work promises to add welcome depth to the still persistent perception (in some quarters) that broadcasting history is monolithically 'institutional'.

Features, Cultures, Modernisms

The energy and experimentalism of the BBC Features Department in itself gives the lie to any assertion that the BBC, as a site of cultural production, was in any way monolithic; histories of its output need therefore to be similarly flexible, nuanced and grounded in its multifarious cultural contexts and intermedial relationships. Of course, what is often referred to as the 'Features Department', referring mainly to its post-war incarnation, did not spring fully formed from the corridors of Broadcasting House. As early as 1928, the *Radio Times* used a full-page article to introduce its readers to the feature programme: a mixture of music, drama and talk, constituting 'an original form of expression, peculiar to broadcasting'.[1] In the 1920s, such features were a result of artistic experiments in the direction of producing *radiogenic* drama, as distinct from dramas that were to an extent derivative of the stage. As a result of this formal experimentation, there was, as producer Rayner Heppenstall noted, 'no short answer to the much-asked question: "What *is* a feature programme?"'; he himself came up with perhaps the best answer by characterising features as 'anything put out by a producer in Features Department'.[2] It is important to note that although production staff were proud to stake a claim to the feature as 'purely a BBC invention',[3] other national broadcasters were similarly engaged in pushing the boundaries of radiogenic form (compare, for example, the tradition of German *Horspiel*).

The generic boundaries of the feature form are fluid and generously adaptable to the meaning a writer and producer intended to be communicated through sound. Practically speaking, features utilise imaginative combinations of speech, drama, music, location recording and sound effects in order to suggest meaning, drawing selectively on, for example, the entire range of possible speech forms and arranging them artfully to prick listeners' imaginations. Features may look, on the face of it, like documentaries, radio plays, radiophonic engagement with modernist poetry, wartime propaganda pieces, travelogues, social history, journalism, etc, but their modes of communication are protean. In the years before the sustained use of actuality and stereo sound became commonplace, for example, many features contained little location recording and sat more within the parameters of dramatic convention rather than in a format comprised of narrated links and contributor clips intercut with sound. The vernacular of features in the time period covered by this volume created rich imaginative worlds by embracing the fluidity of the in-between, interstitial, territory inhabited by the genre itself—one that straddled the boundaries of realism and drama, fact and fiction.

The longstanding significance of the creative innovation of early feature writers including Tyrone Guthrie, Lance Sieveking and Mary Hope Allen is acknowledged by the post-war head of features Laurence Gilliam in his 1950 book *BBC Features*. He discusses the important origins of the Features Department in the form of the Research Unit which in the 1920s explored the dramatic and artistic possibilities of the medium; that unit developed into the Features Unit which, together with the Drama Unit, formed a department under the leadership of Val Gielgud. After the war, the Features Unit was granted the autonomy of a Department, which was led by Laurence Gilliam until his death in 1964, soon after which it was disbanded. Institutional origin myths reveal only so much, however; perhaps features are best understood as products of creative encounters within 'contact zones' made possible by the BBC, 'where cultures, meet, clash and grapple with each other'.[4] These 'cultures'—plural—include artistic, social and political dimensions that are entangled in multiple ways with those of modernism.

The title of this collection also seeks to embrace 'radio modernisms', plural: it champions the expansive conception of modernism which includes the people and the words of the early twentieth-century literary movement but also radio as a technology, a site of cultural production and impactful on the individual's experience of daily life.[5] Radio in itself is multiply expressive of many modernities that can be examined through a number of lenses which relate to perceptions and experiences of individual and collective identities (e.g. race, class and gender). The expansive conception is also open to the inherent intermediality of artistic and cultural practices. We adhere to the temporal extension of modernism beyond the Second World War and into the 1950s—the so-called 'golden age' of radio when many mid-century writers were deeply engaged with experimentation within broadcast forms. In 1952, Gilliam wrote that 'the flow of new imaginative creation for radio shows no sign of drying up. It is a heartening thing to find this, among the youngest of art-forms, attracting the creative effort of some of the best of our contemporary writers'.[6]

In navigating the cross-currents of interest amongst scholars from different disciplines, we have sought to ensure that each author's contribution to this collection has the best chance to speak meaningfully beyond disciplinary boundaries—that is, where possible, engaging with a programme in terms of sound, as a cultural event from a specific moment in time, as team creation, as something heard in its original domestic context and as a cultural work that may have attracted a certain canonical pedigree. This combination of the particular, the contingent and the contextual extends through this special issue.

Kate Lacey's opening essay considers the printed schedule for pre-WWII radio broadcasts (published in the *Radio Times*) to be a kind of modernist text in itself. Its fragmentary fluidity ('the national stream of consciousness') is said to capture and mediate the contradictions of modernity—and for all, with listeners imagined as 'co-authors', enabled to transcend class, gender and social stratifications by the progressive potential of the medium. Todd Avery locates echoes of nineteenth-century aestheticism in the artistic innovation of writer-producers of early radiogenic dramas such as Lance Sieveking and Tyrone Guthrie whose experiments in drama from the 1920s established a rich and welcoming cultural environment for modernist writers such as Beckett, Joyce and Woolf.

Alex Goody moves the discussion into another formative period for the evolution of features when, during the Second World War, the BBC engaged writers to fashion features to counteract Nazi propaganda. Goody explores the moral questions arising for BBC staff concerning the affective power of the voice when used to stand for historical figures in representations of real events in wartime features. Leonie Thomas explores how the poetic voice of the Jamaican poet Una Marson became more experimental as a result of working within the currents of transnational modernism at the BBC, in the heart of the British empire, from 1939 to 1946. Her experience of racism, however, limited her broadcasting career at the pivotal point when—with *Caribbean Voices*—she had helped to construct a platform for other Caribbean writers. Alex Lawrie evaluates the evidence for listeners' engagement with modernist innovation and aesthetics in literary radio programmes broadcast in the 1940s and 1950s, with case studies including a talk seeking to demystify Virginia Woolf, radio presentations of David Jones' high modernist epic poem *In Parenthesis* and Herbert Read's verse play *Moon's Farm*. Certain paths across Louis MacNeice's work as poet, radio writer and producer have been well traced, but Aasiya Lodhi explores new territory in her discussion of how the geomodernist dimensions of several of his boundary-crossing travel features signify the 'reshaped global order' at the moment when Britain, having moved through the Second World War, faced the disintegration of its empire.

E. Charlotte Stevens and John Wyver investigate the intermedial extension of aspects of the radio feature genre into the poetic mode of television documentaries made by Denis Mitchell and Philip Donnellan in the 1950s and 1960s, drawing on their prior experience working on radio features. Stevens and Wyver argue that these producers were able to transfer and transform distinctive aesthetics of 'pure radio' into highly effective visual forms. Also on an intermedial theme, Amanda Wrigley's essay charts the rich afterlives of what are perceived to be 'classic' literary features, such as Edward Sackville-West's *The Rescue*, Louis MacNeice's *The Dark Tower* and Dylan Thomas' *Under Milk Wood*, considering the significance of repeat performances on the radio, realizations in other performance media and, especially, their adaptation into print form in the construction of a sense of a 'canon' of radio features. We are indebted to David Hendy for drawing together some of the pertinent strands in this collection and casting them outwards for scholars of all stripes to interweave.[7]

Disclosure Statement

No potential conflict of interest was reported by the authors.

Notes

1. B. E. N., "Feature Programmes".
2. Heppenstall, *Portrait*, 26, 27.
3. Bridson, *Prospero*, 80.
4. Ranasinha, "South Asian Broadcasters," 62, who borrows from Mary Louise Pratt's usage of the term as a space 'where cultures, meet, clash and grapple with each other, often in contexts of highly asymmetrical relations of power, such as colonialism, slavery, or their aftermath' (34).
5. *Pace* Friedman ("Definitional Excursions," 513) who warns that the plural may suggest heterogeneity without always delivering it.
6. Gilliam, "The Radio Feature," 11. On 'mid-century modernism' as the most appropriate term with regard to modernist aesthetics in media history, see Bloom, *The Wireless Past*, 162–3.
7. We offer warm thanks to Jamie Medhurst and his editorial team at *Media History*, including Stephanie Jones, who have been patient, flexible and unfailingly supportive.

ORCID

Amanda Wrigley http://orcid.org/0000-0003-3822-7633

Bibliography

Addyman, D., M. Feldman, and E. Tonning, eds. *Samuel Beckett and BBC Radio: A Reassessment*. New York: Palgrave Macmillan, 2017.

Avery, T. *Radio Modernism: Literature, Ethics, and the BBC, 1922-1938*. Aldershot: Ashgate, 2006.

B. E. N. "Feature Programmes." *Radio Times*, August 31, 1928, 367.

Bloom, E. C. *The Wireless Past: Anglo-Irish Writers and the BBC, 1931-1968*. Oxford: Oxford University Press, 2016.

Bridson, D. G. *Prospero and Ariel: The Rise and Fall of Radio. A Personal Recollection*. London: Gollancz, 1971.

Cohen, D. R., and M. Coyle, eds. *Broadcast Traces/Tracing Broadcasting: Modernism and Radio*. A Special Issue of *Modernist Cultures* 10, no. 1, 2015.

Cohen, D. R., M. Coyle, and J. Lewty, eds. *Broadcasting Modernism*. Gainesville: University Press of Florida, 2009.

Dinsman, M. *Modernism at the Microphone: Radio, Propaganda, and Literary Aesthetics during World War II*. London: Bloomsbury, 2015.

Drakakis, J., ed. *British Radio Drama*. Cambridge: Cambridge University Press, 1981.

Feldman, M., E. Tonning, and H. Mead, eds. *Broadcasting in the Modernist Era*. London: Bloomsbury, 2014.

Friedman, S. S. "Definitional Excursions: The Meanings of Modern/Modernity/Modernism." *Modernism/modernity* 8, no. 3 (2001): 493–513.

Gilliam, L. "Introduction." In *BBC Features*, edited by Laurence Gilliam, 9–14. London: BBC, 1950.

Gilliam, L. "The Radio Feature." *Radio Times*, September 26, 1952.

Hajkowski, T. *The BBC and National Identity in Britain, 1922-53*. Manchester: Manchester University Press, 2010.

Hansen, M. "The Mass Production of the Senses: Classical Cinema as Vernacular Modernism." *Modernism/modernity* 6, no. 2 (1999): 59–77.

Hendy, D. "Painting with Sound: The Kaleidoscopic World of Lance Sieveking, a British Radio Modernist." *Twentieth Century British History* 24, no. 2 (2013): 169–200.

Heppenstall, R. *Portrait of the Artist as a Professional Man*. London: Owen, 1969.

Hilmes, M. *Network Nations: A Transnational History of British and American Broadcasting*. New York: Routledge, 2011.

Lacey, K. *Listening Publics: The Politics and Experience of Listening in the Public Age*. Cambridge: Polity, 2013.

LeMahieu, D. L. *A Culture for Democracy: Mass Communications and the Cultivated Mind in Britain between the Wars*. Oxford: Clarendon, 1988.

Lewis, P., ed. *Radio Drama*. London: Longman, 1981.

Murphy, K. *Behind the Wireless: A History of Early Women at the BBC*. Basingstoke: Palgrave Macmillan, 2016.

O'Sullivan, J. "Modernist Intermediality: The False Dichotomy between High Modernism and Mass Culture." *English Studies* 98, no. 3 (2017): 283–309.

Potter, S. *Broadcasting Empire: The BBC and the British World, 1922-1970*. Oxford: Oxford University Press, 2012.

Pratt, M. L. "Arts of the Contact Zone." *Profession* (1991): 33–40.

Ranasinha, R. "South Asian Broadcasters in Britain and the BBC: Talking to India (1941-1943)." *South Asian Diaspora* 2, no. 1 (2010): 57–71.

Rodger, I. *Radio Drama*. London: Macmillan, 1982.

Scannell, P., and D. Cardiff. *A Social History of British Broadcasting. Volume One, 1922-1939: Serving the Nation*. Oxford: Blackwell, 1991.

Webb, A. *London Calling: Britain, the BBC World Service and the Cold War*. London: Bloomsbury, 2014.

Whitehead, K. *The Third Programme: A Literary History*. Oxford: Clarendon, 1989.

Whittington, I. "Radio Studies and 20th-Century Literature: Ethics, Aesthetics, and Remediation." *Literature Compass* 11, no. 9 (2014): 634–648.

Whittington, I. *Writing the Radio War: Literature, Politics and the BBC, 1939-1945*. Edinburgh: Edinburgh University Press, 2018.

Wrigley, A. *Greece on Air: Engagements with Ancient Greece on BBC Radio, 1920-1960s*. Oxford: Oxford University Press, 2015.

RADIO'S VERNACULAR MODERNISM: THE SCHEDULE AS MODERNIST TEXT

Kate Lacey

This article explores the modernism of pre-war radio in terms of the framing device of the schedule, rather than exceptional texts or 'features'. It suggests the flow of broadcasting could be experienced as a montage of remediations that invited reflexive engagement with the conditions and contradictions of modernity. The schedule thus appears not only as a site for the mediation of modern experience and a new sensorium, but as a modernist text in its own right that produced, and was expressive of, a pervasive and insistent vernacular modernism.

Introduction

One of the most celebrated pioneers of the BBC radio feature, Lance Sieveking, once described his experimental soundworks as 'themed sounds without a plot'.[1] Such modernist radio montages evidently stood apart from the everyday narratives, both fictional and factual, that were routinely broadcast over the airwaves. But the idea of 'themed sounds without a plot' also unwittingly describes the montage of elements that made up the early BBC's mixed programming schedules.

Many of the familiar tropes of the modernist aesthetic—the fragmentary and the fluid, the dissonant and the distracted, the automatic and the alienated—can be identified in the early scheduling practices when all radio was still, by definition, experimental. Rather than focusing on the exceptional modernist texts that punctured the scheduled flow of everyday radio, and which have been relatively well attended to by critics versed in the techniques of close reading (and even, sometimes, close listening), this article examines the framing device of the schedule itself. The central hypothesis is that the schedule is not only a significant site for the mediation of modern experience and a new sensorium, but that it can be read as a modernist 'text' in its own right. Reading the schedule in this way reveals it as a text that produced, and was expressive of, a pervasive and insistent vernacular modernism; a flow of broadcasting that could be experienced by listeners as a montage of remediations and that acted as an invitation to reflexive engagement with the conditions and contradictions of modernity.

Broadcasting as Vernacular Modernism

Modernism has long been understood as more than a singular aesthetic movement or repertoire of styles associated with a particular canon of avant-garde artists.[2] It is now widely taken to be a term that embraces a whole range of cultural responses to the processes of modernisation and experiences of modernity. The radio is doubly articulated in this modernist moment, being itself both a textual practice and part of the wider context of changing communicative practices and possibilities; a site, in other words, for new forms of artistic expression, a new sensory regime and a new politics of consumption.

But beyond this familiar duality of text and context, I want to argue that the form that British broadcasting assumed in its first decade or so—broadly speaking, a schedule of mixed programming—can also be read as a site for what Miriam Bratu Hansen, in relation to classic Hollywood cinema, called 'vernacular modernism'.[3] Hansen used the term to describe American movies as a popular mass cultural form that was contemporary with high modernist experimentation, but which had long been widely—albeit problematically—conceptualised as its other.[4] If Modernism offered a reflexive and critical engagement with the dislocations of modernisation and a rejection of classical aesthetic traditions, Hansen argued that Hollywood films—produced for (and producing of) a new mass public via the most modern technological means of the day—correspondingly provided a way of translating (that is to say, mediating and expressing) the contradictions of modernity into everyday experience. She explained her preference for the term 'vernacular' over 'popular' in that it avoids loaded assumptions and value judgements, while carrying with it the idea of everyday usage as well as 'connotations of discourse, idiom, and dialect, and with circulation, promiscuity, and translatability'.[5] She was interested less in on-screen modernist techniques than in the way that, 'even the most ordinary commercial films were involved in producing a new sensory culture'.[6]

Vernacular modernism in this sense is a term full of resonance for early broadcasting. Radio was certainly involved in a mass public—and, at times, popular—address; and it was involved in the production of a new sensory culture: one based on the simultaneous encounter of distant listeners with dislocated and disembodied sounds. It was certainly vernacular—both in the sense of bringing the spoken word back into public life (albeit, in the case of the BBC, a form of speech that was in the early years predominantly scripted, and articulated in the tones of received pronunciation), and in the sense of being woven into the warp and weft of everyday life in communal, and especially domestic, spaces. Indeed, it is arguably in its domestication and everydayness, as much as in its lack of a visual image, that broadcasting most clearly differs from the glamour and spectacle of the cinema. But, as I will argue, the form of the schedule and the domesticated and distracted conditions of its reception make it no less important a site for the mediation and translation of modernity's contradictions.

There has certainly been a recent flowering of interest in radio modernism in Modernist Studies, a field long dominated by the visual, musical and literary arts. But, broadly speaking, this interest has tended to focus on certain modernist *auteurs*, or the transmission of particular instances of modernist theatre, music or poetry. Some address the idea of the 'radiophonic', others the phenomenology of listening to modernist texts on the air.[7] Todd Avery (next in this volume), has shown how the ethos of public service broadcasting itself influenced the work of modernists working in other arts in the interwar period.[8] And of course there has been plenty of attention given to the role of radio as a modernising influence in general.[9] Some writers have also touched on radio's vernacular modernism. Aaron Jaffe identifies it at the moment of radio's technical inception when a modernist 'rhetoric of invention' was translated into the foundational and popularising myths of the new technology.[10] Shawn VanCour, meanwhile, invokes the term in his discussion of mobile, ambient and distracted regimes of modernist radio reception.[11] My contention, however, is that the broadcast schedule itself—the very 'stuff' of radio, to adapt Sieveking's phrase—is marked by modernist tropes and techniques, making it the quintessential vehicle for radio's contribution to the vernacular modernism of the pre-war period.

Conceiving the Schedule

Conceptually, the schedule lies somewhere between text and medium, between form and formlessness. It is the institutionalised response to the technical capacity of continuous transmission. It is in the act of scheduling that the transmission of discrete sounds and voices over radio waves is metamorphosed into broadcast form. The schedule is both the thing listened to and the silent framing of that which is listened to. It comprises all that can be heard while at the same time being something that always already exceeds that which can be listened to by any individual. It is the institutionalised address to a listening public which is itself only produced in the moment of its being addressed. In its written form the schedule exists in anticipation of sounds as yet unsounded, and survives as archived record of the live and ephemeral sonic texts of this period that are lost to history. It represents the whole of the broadcast output in representing the fragments of programming that make up its flow. It follows predictable patterns but it is read—in advance by audiences, and by historians in retrospect—primarily for its marking out of the individual instances of any particular day. It is, in short, not merely the context within which the modernist radio text appears, but is itself a complex and contradictory modernist phenomenon.

The schedule is a text that is not a text. Historians of early radio have long grappled with the almost complete absence of a sound archive, and instead negotiate its graphic traces left behind in the written archives, the contemporary commentaries and the listings magazines.[12] These anticipatory and reconstituted representations must remain, however, only partial witnesses to the schedule as it manifested in sound. Then again, the schedule in sound was always already something of a fiction, beyond the apperception of any normal listener (listeners tend to listen intermittently), a sum always somehow less than its programmed parts. Though the schedule might be co-extensive with all that is broadcast, we never speak of listening to a schedule. We listen to 'the radio' or to particular named programmes, stations or shows; the schedule as listening event disappears in the acts of reception and recall. There is a strange disavowal at work here, even in those moments when the announcer speaks, assuming the role of the meta-narrator, giving voice to the schedule. The invisible and unnamed announcer ventriloquises the institutional framing of the other-authored texts.[13] On a national wavelength the ubiquitous and authoritative voice of the anonymous announcer might even stand in, at some imaginary (ideological) level, for the voice of the nation itself.[14] But inasmuch as the schedule is the disembodied voice of the institution of broadcasting, it also exemplifies and amplifies the disembodiedness, the deep uncanniness, of all radio voices. It amplifies this disembodiedness in the sense that this particular voicing body, this 'voice of the BBC', is encoded as anonymous, depersonalised and interchangeable.[15]

Early Scheduling by the BBC

A week before the British Broadcasting Company was originally due to begin transmissions in the summer of 1922, the *Times* announced that it was expected 'to provide a six hours' programme every evening from 5-11', apart from Sundays, 'when the programme will occupy practically the whole day'.[16] Although there was some brief mention of the intended content, including non-denominational sermons and 'dead' (already published) news, it is striking how a story about the imminent launch of a long-anticipated new medium leads on the time to be filled, on the empty premonition of the schedule.

The first day's broadcasting, on 14 November 1922, consisted of short bulletins of news and weather, read out twice at different speeds by the Programme Director, Arthur Burrows, with a request to listeners to let him know which they preferred.[17] However, it wasn't long before regular schedules developed that included a daily and diverse diet of music and drama, lectures and literature, sermons and sport. Yet the early BBC had a famously ambivalent attitude to the idea of a fixed schedule. On the one hand, the institution needed to plan and transmit material daily in sequence. On the other, there was a resistance to producing a predictable or seamless flow of programming that might encourage profligate and indiscriminate listening.[18] Listeners were enjoined to study published schedules in advance, in order to focus their listening purposively on individual elements. To this extent, there was a performative disavowal of the schedule by those who produced it.[19]

Another way to read this is as a perverted sort of Brechtian *Verfremdungseffekt*.[20] Here the construction and framing device is in plain and deliberate view, but the audience's attention is deliberately directed elsewhere. The listings serve as a reminder of the contingency and constructedness of the form, and yet fade away in the act of listening. In this sense, the schedule is emblematic of the experience of all radio as frameless or immaterial: the radio apparatus is forgotten in the moment of absorption in the ephemeral radio text, but also in the moment of distraction when used as background noise or sonic companionship. The radio set is rarely the object of the gaze; indeed, its domestication for mass consumption as the decade rolled on involved disguising its technicity as mere furnishing. But even if this furnishing does become an object of design and display, when in use it does not need to be looked at.[21] In that moment, the listener becomes the receiver. The stuff of radio in all its ephemerality, invisibility and ethereality transcends and overwhelms the materiality of the radio set. But the listener-as-receiver is at the same time a co-author, a producer of mental images and associations to accompany invisible sounds. Radio's intimacy and immediacy, then, afforded new forms of interactivity and engagement. All of this is by now well known, second nature; but once it was disruptive, novel, modern.

By extension, even the most conventional content could be rendered novel and modern in its translation into this virtual, wireless world. Much of the content and sensibility might have been positively Victorian, but it bore of necessity the imprint of technology, science and innovation. At the same time, since most material at this time was poached from prevailing cultural tastes and traditions, it is hardly surprising that early radio bore the birthmarks of a deeply stratified society. A 'general public' was constituted by the technical indiscriminacy of broadcasting, but in practice that public was very often addressed sequentially in terms of various—though far from exhaustive—imagined constituencies.

The first weekday schedule listed in the *Radio Times* (for the London station, 2LO, on Monday, 1 October 1923) illustrates the point: the broadcast day began at 11.30 with the Morning Concert (classical, of course); then a break until *Women's Hour* at 17.00, *Children's Stories* at 17.30, *Boys' Brigade and Boys' Life Brigade News,* followed by a 35-minute 'interval' until the *1st General News Bulletin* at 19.00, the short *Weekly Book Talk* at 19.15 and a *Symphony Concert* at 19.30. There then came a talk about aeroplanes by someone from the Air Ministry at 21.10, the *2nd General News* at 21.30 and a final classical concert from 21.45 until *Closedown* at 22.30. The rest of the weekday output followed a similar pattern with different youth groups' news, talks on theatre and films, and the inclusion of some dance music towards the end of the week.

Although it is true that all these various listening constituencies were imagined through a thoroughly middle-class filter, and can be understood as connected to modern disciplinary techniques of division and categorisation, it is nevertheless also true that within the course of a single programmed day the schedule might address any singular listener at turns *as if* they were a woman, man or child, a reader, concert-goer, student or citizen. While individual shows ostensibly spoke to an idealised and specialised listener, the schedule obliquely invited any individual listener to adopt a series of guises—to masquerade repeatedly in borrowed identities, or inhabit the role of interloper or eavesdropper. The polymorphic form of the schedule in this emergent moment thus interpellates a fractured and fluid subject, a listener who is marked by the modern, able to traverse longstanding social and cultural divides and overcome distances in space and time and class.

Mixed Messages

Over the years, in both public service and commercial contexts, the 'art' of scheduling has developed with the precise aim of maximising the size of the audience, either by matching the daily (gendered) routines of domestic and mobile audiences, or by targeting ever more precisely defined taste publics. By 1928 the BBC was already beginning to offer alternative programming via simultaneous broadcasts on different wavelengths. At this point it began to conceptualise its programming as either 'universal' or 'specialist': the former demanded no particular concentration, whereas the latter would require deliberate and focused attention.[22] One way to understand the distinction is to recognise this as a moment where the institution begins to loosen its attachment to pre-existing cultural practices and begins to come to terms with the radiogenic character of broadcast flow and the realities of secondary listening.

The standardisation of the schedule in terms of daily output, time slot or duration of particular programmes was a gradual affair in the BBC's first decade.[23] Sometimes content had not been finalised by the time the *Radio Times* went to press. This was the case, for example, with the series of 'Special Features' broadcast on Wednesday evenings on 2LO, the London station, throughout the summer of 1925, the details of which would 'be announced through the microphone, and will appear in the Daily Press.'[24] Asa Briggs describes the struggle to define any sort of principle for scheduling during this early period beyond two broad principles: 'contrast' (often delivered by the alternation of music and the spoken word), and 'balance'. This was a policy driven in part by an anxious desire to ensure that no one listener was alienated for too long by finding nothing to their taste.[25] Eventually this came to produce a 'middlebrow' schedule defined, in Toby Miller's withering phrase, by 'the law of maximum inoffensiveness'.[26] But in these pioneering years, the bricolage of broadcasting was not yet fully tamed.

The schedule, already a long way from broadcasting all day long,[27] was further punctuated by lengthy periods of silence to signal the output as a series of discrete cultural productions. The intention was to encourage listeners to prepare for, or reflect on, their listening, with just a ticking clock to reassure the audience such breaks were not technological breakdowns.[28] In this way, the schedule tapped into a deep modernist concern with questions of connectivity and its failures. But for Reith, the mixed schedule with its distinct elements and moments of silence was a key instrument in the moralising project of self improvement and social uplift.[29] The resistance to standardisation in principle allowed flexibility for producers, while enabling chance encounters with unfamiliar or improving fare for listeners.[30] It is just one of the constitutive paradoxes of these early schedules that the content, so often shaped by longstanding traditions of value judgement, was presented within such a radically novel frame.

The Schedule as Automatic Writing

One of the novelties of Modernism is its fascination with automatism as a route to the unconscious. The preoccupation of many writers and artists with the psychic, the spiritual and the dreamlike, gave rise to a series of techniques and styles of creative production that tried to circumvent tradition and authority or to liberate consciousness in some way. Technologies old and new were invoked in these movements variously as prosthetic conduits to new states of being and subjectivity, or as instantiations of an amputational and dehumanising machine logic.[31] If the chosen technology of the surrealists in their classic experiments with automatic writing was the pen, it was only as a tool of 'magic dictation' as artists tuned into an unconscious interior monologue 'akin to spoken thought', producing a rapid trace with 'no clear beginning or end, and (seemingly) a lack of narrative and of style'.[32]

It is evident that broadcast radio was at first an ersatz medium, a substitute for theatres, cinemas, newspapers and other places of public congregation and communication. In other words, the early schedules were characterised by the logic of remediation: the repurposing of already available plays and concerts and newspaper reports, children's stories, church sermons and academic lectures; the reviews of films and books; the relay of ceremonies and sporting occasions. Such a list also reveals a high degree of intermediality —radio was immediately imbricated in wider cultural networks, its producers borrowing and refashioning both form and content from sources across the realms of information, entertainment, education and the arts. The schedule was promiscuous in its attachments, albeit these relationships were not always straightforward nor uncontentious.[33] The schedule also regenerated itself daily, year after year. It therefore reflected, with varying degrees of selectivity and distortion, the unfolding history and preoccupations of the age.

It is tempting, then, to read the schedule as a form of automatic writing at the level of an institution—humanly produced, yet impersonal, machinic and abstracted. The schedule represents the long—potentially never-ending—form of radio; authored, yet somehow also authorless. Indeed, over time, and seen from enough of a distance, the relentless flow, drawn from all the tributaries of journalism, literature, music and science, melds into something like the nation's stream of consciousness, its interior monologue with itself (or at least that hegemonic form of the national culture that made its way to air).

Of course none of these borrowed genres were conveyed unchanged, and the historical schedule can reveal the gradual emergence of specifically broadcast genres, from studio discussion to soap opera to sporting commentary—not to mention the celebrated radio 'feature' dedicated to exploring the possibilities of the radio form.

Features, Fragments and Flaneurs

Of course, the feature was one of the key locations for modernist experimentation in the early BBC. But looking at the way that the term 'feature' featured in early schedules indicates it took time for it to bed down in the way that now can seem inevitable. The *Radio Times* listings show the term 'feature' could mean anything from music written for radio to recitals, plays, humorous interludes and specials on subjects like football. Some interactive experiments also catch the eye—for example, 'A New Feature by Old Friends', broadcast by 2LO on Saturday afternoon, 14 March 1925. Listeners were invited to suggest a title after listening in, the best suggestion winning the chance to attend the next feature performance in the studio. Or 'Puzzle Programme', broadcast on 23 November that year by 6BM (Bournemouth), which invited listeners to complete the blanks in a short story published in the magazine, using words found from the items played by the orchestra during the programme. Some flew obstinately in the face of the lack of visuals, like the weekly feature, 'Drawing by Numbers' with Heath Robinson. Others still were labelled as 'experimental transmissions', either as a space for 'amateurs', or sometimes for the transmission of sound and vision—on separate wavelengths—of nascent television broadcasts. Of course, these few illustrations are themselves just residues picked from the schedules, highlighted here to nod to the diversity and experimentation above and beyond the classic modernist feature, and to highlight the makeshift construction of the early schedule from a series of fragments.

In pre-broadcasting radio days, the dominant listener experience had been of fishing for fragments of sound from the white noise of the boundless ether. Wireless content mattered less than wireless contact.[34] If radio reception was born in this random and serendipitous realm, then early scheduling practices can be understood as the gradual domestication and cohesion of this fragmented experience. Eventually this insistent logic would see a widespread shift from mixed schedules through specialist and genre-specific channels to the personalised programming of so-called smart radio today. But early BBC schedules were not yet shaped by these logics of commodification and pseudo-individuation. Mixed programming assembled the flow of time into discrete fragments.

Raymond Williams identified such fragments as 'timed units', alluding to broadcasting's part in the broader industrialisation of culture.[35] From this perspective, the schedule is the mechanism by which broadcasting contributes to the modern rationalisation and commodification of time, and which acts as the conduit for the shared and simultaneous experience over distance that is the hallmark of the modern mediated condition. Williams, in his influential discussion of broadcast flow, makes the argument for radio programming marking a break from all previous forms of cultural arrangements, and the 'specific, isolated and temporary' forms of attention associated with them. At the same time, though, he recognised how this mixed programming was an expression of the increasing 'variability and miscellaneity of public communications', itself a response to profound changes in the levels of social and physical mobility, and the privileging of consumer over community culture. He noted how the radio moved from the transmission of events already happening in the world to how the demands of the schedule and the power of the institution began proactively to *programme* those events. Here we see in the logic and power of the radio schedule the inception of the processes of 'mediatization'.[36] Eventually, the logic of continuous output to capture the audience would produce, as Williams described it, a flow without interval: a 'generally available experience' that has by now become the quotidian experience of media.[37]

The idea of the fragment also holds a hallowed place in the aesthetic modernist imagination, revealing now the disjointedness and disembeddedness of the modern experience, then a shoring against the ruins.[38] At the risk of straining an analogy, the fragmentary writing of the schedule shares with its literary counterpart what Maurice Blanchot described as its 'distinguishing characteristic', namely 'the interruption of the incessant [...] interruptions having somehow the same meaning as that which does not cease'.[39] This is not just to recognise the interruption of the rounds and routines of private life with the rhythms and rhetoric of public life. Nor is it just to reassert the segmentation of the broadcast flow into the disparate programming formats. Rather, it also speaks to the constant repetition without reproduction, the familiar framing of constant novelty, the refusal of any real sense of final closure as the daily, weekly, seasonal and annual round promises ever more of the almost same. Just as the fragments from different realms and times of cultural life produced the sense that 'radio' exists out of time, or in all times simultaneously, so the mix of voices suggest at least the possibility of a space beyond a single locus of power. The relentless unfolding of the schedule over years and decades produces through these fragments a communicative text that exceeds comprehension or control.[40]

From the start, then, the BBC developed a mixed programming style, mixing up in a single channel music and the spoken word, the serious and the popular, the factual and the fictional, the general and the bespoke, the traditional and the modern. This daily mosaic was complicated by the weekly and seasonal rhythms of the schedule as it echoed and amplified a national calendar.[41] The result was a kind of polyvocal and polyphonic bricolage. The printed schedules provided a map that charted a linear, unified route through the aural collage; but the repeated exhortations to use the map to isolate features in that acoustic landscape suggest that listeners were perhaps more likely to wander aimlessly than head for specific destinations.[42] Here we might find the figure of the listener as flaneur—not the 'strolling detached observer' of the city streets and byways, but the distracted, detached auditor of the airwaves.[43]

The Shock Effect of the Schedule

If the institution had to learn the art of scheduling, then so too did listeners have to adapt. Although radio technology had been around since the turn of the century, the experience of hearing voices over the airwaves only gradually became part of everyday experience. There are plenty of accounts through the 1920s of the wonder of radio reception; discourses of magic and mystery jostle alongside scientific and technical explanations.[44] But the shock effect of hearing disembodied and distant voices was compounded by exposure to voices that were simply unfamiliar. While the few cosmopolitan 'early adopters' might have closely matched the small middle-class coterie of voices on the airwaves, with time the variety of accents and interests began to widen, producing spaces for entirely new and often unexpected encounters. If listeners did not at first know how to react to the voices of monarchs and ministers in their parlours—witness establishment anxieties about listeners acting inappropriately while in the acoustic presence of these absent figures[45]—neither did they necessarily know how to react to the orchestration of mannerisms and modes of address appealing for their attention. This active and continuous experience of displacement and dissonance in daily life is surely one of the key devices of radio's vernacular modernism.

It was in part the unknowability of the absent audience that shaped so much of the BBC's thinking about the schedule. Apart from its initial resistance to audience research (for fear it could be misdirected into giving this phantasmatic audience what they wanted, as opposed to what was good for them), the BBC was decidedly anxious about the apparent power of this untested medium in an age when 'the masses' were widely conceived of as both volatile and suggestible, where domestic audiences that included women and children were understood as innocent and vulnerable, and where the power of the spoken word was being unleashed on an unprecedented scale in the public sphere.[46] All this, set against the prevailing anxieties and uncertainties of an age caught between the aftermath of one war and the anticipation of another[47]—anxieties and uncertainties that were so often the impetus and subject of modernist treatments—helped produce a schedule that was constrained by convention and routinely steered to the middle of the road.

And yet here is another of the express contradictions of the BBC in relation to its schedule-building. Though listeners were enjoined to listen with discernment and intent, presumably to programmes that already 'spoke to them' in some way, so too was it an explicit policy to enable listeners to chance upon improving or challenging material by scheduling it between other attractions. Would it be too much of a leap to think about the dysjunctures between sequential programmes of different styles and voices like jump cuts and film frames having the potential to induce a sort of shock effect—that celebrated modernist technique to jolt people out of their comfort zones?[48]

The intimacy of radio's address—of human voices speaking to listeners in their own homes—reinforces this invitation. The fact that a listener's own domestic soundscape was overlain with the artifice of sounds curated elsewhere breathed the air of modernity right into the home and into everyday life. I have argued elsewhere about the particular import this had for women working in the home who were otherwise, in this period following the First World War, the least integrated into the modern.[49] Women like these were not just newly exposed to ideas and voices in the public sphere, and were not just newly connected imaginatively to other isolated women in domestic spaces; they were also directly involved in this strange new fluid and fragmentary sensory regime that the mixed broadcasting schedule relayed into their everyday lives. That the vernacular modernism of the schedule, produced in and expressive of the changing public sphere, was matched to the archetypal rhythms of the family day had a progressive potential to make manifest that which is by definition unremarked and unremarkable: the monotony, repetitiveness, interiority and interruptibility of the everyday.[50]

Conclusion

Early BBC scheduling exists in a strange place between intention and accident.[51] It is the product of not yet institutionalised practices within the institution. It is the novel result of juggling the different ideas and contributions of the various producers and departments working out what it means to have to fill a permanently available stage before a transient and invisible audience. It is the corporate application of an upstart cultural technology striving, by remediating more conventional cultural forms, to find recognition as an arm of the establishment.

There are, then, good reasons to attend to the schedule as something that exceeds the individual text or genre, but that still gives itself up to the techniques and insights of textual analysis. In so doing there is, I would suggest, a strong case to be made on various counts for the early broadcast schedule as modernist text and, more particularly, as a vehicle for vernacular modernism. It has, perhaps, been overlooked as such precisely because it has, over time, taken on the mantle of second nature. The schedule is, after all, the given environment against which named and celebrated 'features' stand out. Perhaps, too, it has been overlooked because the schedule is a text of uncertain authorship—produced in the confluence of input from policymakers, producers and programme makers certainly, but also in the expectations, and experience of its publics; or because it is also a text of uncertain content—unending, polymorphic and contradictory, elastic enough to contain both high modernism and kitsch.

Despite its ill-defined authorship, its hybridity and its transient content, the early mixed programming schedule is not a text without purpose, import or meaning. Rather, it represents (in both senses of that word), the logic of broadcasting practice; an instance of 'intentionality without intent'.[52] The schedule is the medium's self-consciousness made manifest, in the sense that it is through the schedules that the institution of broadcasting speaks itself, revealing its purpose, its patterns, its predilections, its principles and its publicness.

In terms of form, too, the schedule can be argued to share in many of the tropes and techniques that so fascinated the modernists, and is matched in its modernity by the conditions and experience of its reception. Such a formalist reading has its limitations in terms of drawing conclusions about either intentions or consequences. Modernism comes in many guises after all, both reactionary and progressive. But such a reading might at least highlight the ambivalences and contradictions within early broadcasting as a whole, complicate some of the easy stereotypes about the early BBC, and help expand the discussion about radio modernism beyond the canon and beyond the limits of the archive.

Disclosure Statement

No potential conflict of interest was reported by the author.

Notes

1. Sieveking, *The Stuff of Radio*, n.p. For Sieveking's contribution to radio modernism, see Hendy, "Painting with Sound".
2. Rainey and Von Hallberg, "Editorial/Introduction," 1.
3. Hansen, "Mass Production".
4. Huyssen summarises this ambivalent relationship in *After the Great Divide*, vii–xii.
5. Hansen, "Mass Production," 60.
6. Ibid., 71.
7. Some indicative titles: Campbell, *Wireless Writing*; Cohen, et al. (eds), *Broadcasting Modernism*; Dinsman, *Modernism at the Microphone*; Porter, Hepburn, "Acoustic Modernism"; Whittington, "Radio Studies and 20th Century Literature".
8. Avery, *Radio Modernism.*
9. Scannell, "Public Service Broadcasting"; Scannell, *Radio, Television.*
10. Jaffe, "Inventing the Radio Cosmopolitan".
11. VanCour, "Early Radio Listening," 8.
12. Historians in many fields are used to dealing with incomplete archives, and Josie Dolan for one has argued that radio historians should not fetishise the sound archive: Dolan, "The Voice that Cannot Be Heard".
13. During this period, the announcer was less likely to be a named personality, or a figure with whom the listener was expected to develop a familiarity, other than as a cypher for the organisation.
14. Mugglestone, "Spoken English and the BBC".
15. Of course, the voice could be decoded otherwise. While the BBC wanted its announcers to be aloof, it seems that, 'An announcer could not cough during a broadcast without receiving presents of everything from cough lozenges to woollen underwear'. http://news.bbc.co.uk/aboutbbcnews/spl/hi/history/noflash/html/1930s.stm
16. "Wireless For Six Hours A Day," *The Times*, 15 August 1922, 5. Negotiations with the Post Office meant transmissions did not, in fact, begin until November that year.
17. BBC, "About BBC News: 14 November 1992, the BBC Takes to the Airwaves," http://news.bbc.co.uk/aboutbbcnews/spl/hi/history/noflash/html/1920s.stm. Note how modernist tropes of experimentation, distanciation and interactivity were already inscribed in this very first broadcast.
18. Scannell and Cardiff, *A Social History*, 372–3; Lacey, "Towards a Periodization"; Goodman, "Distracted Listening".
19. Note this is a world away from later incarnations of the schedule as a vehicle to keep an audience listening across programme breaks, with the use of cliffhangers and explicit exhortations to stay tuned in.
20. The 'alienation effect', or 'distanciation' in Brecht's theatre was a technique to draw the audience's attention to the artifice of a production in the hope of producing a critical distance in place of bourgeois identification.
21. Moores, "The Box on the Dresser", 31–33.
22. Briggs, *The History*, 27.
23. Scannell and Cardiff, *A Social History*, 371; Crisell, *An Introductory History*, 53.
24. Schedule details are drawn from *The Radio Times* listings, available via the BBC Genome project: http://genome.ch.bbc.co.uk

25. Of course there were pragmatic institutional decisions at play too, to do with cost and representation of all the various departments of the BBC. Briggs, *The History*, 30–31.
26. Miller, "An Editorial," 1. See also, Frith, "The Pleasures of the Hearth".
27. Even by 1927, the broadcast day did not start until 10.30am, and then only with the shipping forecast, read twice, the second time at dictation speed, with a long break until schools programming in the mid-afternoon. Briggs, *The History*, 25.
28. Young, *Shall I Listen?*, 256–257; Lacey, *Listening Publics*, 82–83.
29. LeMahieu, "John Reith 1889–1971," 195.
30. Hendy, *Public Service*, 58–59.
31. Armstrong, *Modernism*; Danius, *The Senses of Modernism*.
32. Bauduin, "The 'Continuing Misfortune," 6.
33. See Briggs, *The BBC*, 15; Chignell, *Public Issue Radio*, 13–14; Nicolas, "Media History".
34. Sconce, *Haunted Media*, 65.
35. Williams, *Television*, 88.
36. 'Mediatization' is a term that attempts to encapsulate how pervasive media profoundly reshape (rather than simply reflect or represent) all aspects of contemporary social life. c.f. Couldry and Hepp, "Conceptualizing Mediatization".
37. Ibid., 96.
38. 'These fragments I have shored against my ruins'. Line 430 of T.S. Eliot's *The Waste Land*.
39. Blanchot, *The Writing*, 21.
40. Even with a relatively short broadcasting day by current standards, the BBC broadcast 65,800 hours of 'output' in 1927. Briggs, *The History*, 24.
41. Scannell and Cardiff, ibid.; Hendy, *Radio*, 178. This sonic landscape is further complicated by the patchwork of local, regional, international and 'pirate' broadcasts available to listeners at any one moment.
42. Goodman, "Distracted Listening"; Lacey, *Listening Publics*, 35–37, 128–131, 183–186.
43. VanCour, "Early Radio Listening".
44. Sconce, *Haunted Media*.
45. Scannell, "Public Service Broadcasting," 149.
46. Scannell, "Public Service Broadcasting," 143, "The Box," 23–26; Lacey, 2016.
47. Overy, *The Morbid Age*.
48. Benjamin, "Work of Art". Note these are arguments about latent listening positions rather than speculations about actual audience effects or responses.
49. Lacey, "Towards a Periodisation".
50. Blanchot in Moran, "Criticism," 175; Lacey, *Listening Publics*, 129–130.
51. Scannell, *Radio*, 9.
52. Bourdieu and Wacquant, *An Invitation*, 19.

Bibliography

Armstrong, T. *Modernism, Technology, and the Body: A Cultural Study*. Cambridge University Press, 1998.

Bauduin, T. M. "The 'Continuing Misfortune' of Automatism in Early Surrealism." *Communication +1* 4, no. 10 (2015): 1–42.

Benjamin, W. "The Work of Art in the Age of Mechanical Reproduction." In: *Illuminations*, translated by H. Zohn, 211–244. London: Fontana, 1936/1992.

Blanchot, M. *The Writing of the Disaster*, translated by A. Smock. Lincoln: University of Nebraska Press, 1995.

Bourdieu, P., and L. Wacquant. *An Invitation to Reflexive Sociology*. Cambridge: Polity, 1992.

Briggs, A. *The BBC: The First Fifty Years*. Oxford: Oxford University Press, 1985.

Briggs, A. *The History of Broadcasting in the UK*. Oxford: Oxford University Press, 1995.

Campbell, T. *Wireless Writing in the Age of Marconi*. Minneapolis: University of Minnesota Press, 2006.

Chignell, H. *Public Issue Radio: Talks, News and Current Affairs in the Twentieth Century*. London: Palgrave Macmillan, 2011.

Cohen, D. R., M. Coyle, and J. Lewty. *Broadcasting Modernism*. Gainesville: University Press of Florida, 2009.

Couldry, N., and A. Hepp. "Conceptualizing Mediatization: Contexts, Traditions, Arguments." *Communication Theory* 23, no. 3 (2013): 191–202.

Crisell, A. *An Introductory History of British Broadcasting*. London: Routledge, 1997. Second Edition. London: Routledge, 2002.

Danius, S. *The Senses of Modernism: Technology, Perception, and Aesthetics*. Cornell University Press, 2002.

Dinsman, M. *Modernism at the Microphone: Radio, Propaganda, and Literary Aesthetics During World War II*. London: Bloomsbury, 2015.

Dolan, J. "The Voice that Cannot Be Heard: Radio/Broadcasting and 'the Archive'." *Radio Journal: International Studies in Broadcast & Audio Media* 1, no. 1 (2003): 63–72.

Eliot, T. S. *The Waste Land and Other Poems*. London: Broadview, 2011.

Feldman, M., H. Mead, and E. Tonning, eds. *Broadcasting in the Modernist Era*. London: A&C Black, 2014.

Frith, S. "The Pleasures of the Hearth: The Making of BBC Light Entertainment." In *Music for Pleasure*. Cambridge: Polity Press, 1988.

Goodman, D. "Distracted listening: On Not Making Sound Choices in the 1930s." In *Sound in the Age of Mechanical Reproduction*, edited by D. Suisman and S. Strasser, 15–46. Philadelphia: University of Pennsylvania Press, 2010.

Hansen, M. "The Mass Production of the Senses: Classical Cinema as Vernacular Modernism." *Modernism/Modernity* 6, no. 2 (1999): 59–77.

Hendy, D. "Painting with Sound: The Kaleidoscopic World of Lance Sieveking, a British Radio Modernist." *Twentieth Century British History* 24, no. 2 (2013): 169–200.

Hendy, D. *Public Service Broadcasting*. Basingstoke: Palgrave Macmillan, 2013.

Hendy, D. *Radio in the Global Age*. Cambridge: Polity, 2000.

Hepburn, A. "Acoustic Modernism: BBC Radio and *The Little Girls*." *Textual Practice* 27, no. 1 (2013): 143–162.

Huyssen, A. *After the Great Divide: Modernism, Mass Culture, Postmodernism*. Bloomington and Indianapolis: University of Indiana Press, 1986.

Jaffe, A. "Inventing the Radio Cosmopolitan: Vernacular Modernism at a Standstill." In *Broadcasting Modernism*, edited by D. Cohen, M. Coyle, and J. Lewty, 11–30. Gainesville: University Press of Florida, 2009.

Lacey, K. *Listening Publics: The Politics and Experience of Listening in the Media Age*. Cambridge: Polity, 2013.

Lacey, K. "Listening with Mother: The Cultivation of Children's Radio." In *The Child Savage, 1890–2010: From Comics to Games*, edited by L. Wesseling, 151–170. Farnham: Ashgate, 2016.

Lacey, K. "Towards a Periodization of Listening: Radio and Modern Life." *International Journal of Cultural Studies* 3, no. 2 (2000): 279–288.

LeMahieu, D. L. "John Reith 1889–1971." In *After the Victorians: Private Conscience and Public Duty in Modern Britain*, edited by S. Petersen and P. Mandler, 189–208. London: Routledge, 1994.

Miller, T. "An Editorial Introduction for Radio." *Continuum* 6, no. 1 (1992): 5–13.

Moores, S. "'The Box on the Dresser': Memories of Early Radio and Everyday Life." *Media, Culture and Society* 10, no. 1 (1988): 23–40.

Moran, J. "Criticism: Benjamin and Boredom." *Critical Quarterly* 45, no. 1-2 (2003): 168–181.

Mugglestone, L. "Spoken English and the BBC: In the Beginning." *AAA-Arbeiten aus Anglistik und Amerikanistik* 33, no. 2 (2008): 197–215.

Nicolas, S. "Media History or Media Histories? Re-addressing the History of Mass Media in Inter-war Britain." *Media History* 18, no. 3-4 (2012): 379–394.

Overy, R. *The Morbid Age: Britain Between the Wars*. London: Allen Lane, 2009.

Rainey, L. S., and R. Von Hallberg. "Editorial/Introduction." *Modernism/Modernity* 1, no. 1 (1994): 1–3.

Scannell, P. "Public Service Broadcasting and Modern Public Life." *Media, Culture and Society* 11, no. 2 (1989): 135–166.

Scannell, P. *Radio, Television and Modern Life*. London: Blackwell, 1996.

Scannell, P., and D. Cardiff. *A Social History of British Broadcasting. 1. 1922–1939, Serving the Nation*. London: Blackwell, 1991.

Sconce, J. *Haunted Media: Electronic Presence from Telegraphy to Television*. Durham, NC: Duke University Press, 2000.

Sieveking, L. *The Stuff of Radio*. London: Cassell, 1934.

VanCour, S. "Early Radio Listening as Modernist Practice: Ambient Radio and the Aesthetic of Distraction." *Modernist Cultures* 10, no. 1 (2015): 6–25.

Whittington, I. "Radio Studies and 20th-Century Literature: Ethics, Aesthetics, and Remediation." *Literature Compass* 11, no. 9 (2014): 634–648.

Williams, R. *Television: Technology and Cultural Form*. London: Fontana, 1974. Third Edition. London: Routledge, 2003.

Young, F. *Shall I Listen? Studies in the Adventure and Technique of Broadcasting*. London: Constable, 1933.

WAVES: AESTHETICISM, RADIO DRAMA AND VIRGINIA WOOLF

Todd Avery

The early BBC was devoted to individual and social uplift through the dissemination of 'culture', which the Victorian cultural theorist Matthew Arnold had defined as 'the best that has been thought and said in the world'. Histories of the BBC understandably tend to reduce the impetus to public service broadcasting's mission to inform, educate, and entertain to an Arnoldian influence. Aesthetically and ethically speaking, the situation was more complex. The work of early radio theorists and administrators, and of creative writers and producers such as Tyrone Guthrie and Lance Sieveking, bears within it the traces, sometimes ghostly, sometimes bold, of another, and competing, Victorian discourse—namely, aestheticism, whose roots lie in the work of Arnold's contemporary Walter Pater. Through a closer attention to radio modernism's aestheticist inheritance, this essay also introduces a new phrase into the lexicon of modernist 'moments'—Sieveking's 'moments of hearing'—and shows how the BBC created a space not only for 'moderately adventurous' aesthetic experiments, but also for artistic efforts that were as radically experimental, and as thoroughly in the aestheticist tradition, as works by prominent modern novelists.

The ethos of the early BBC under the leadership of John Reith embodied a morally elevating, socially unifying, paternalistic ambition derived from the cultural theory of the Victorian writer Matthew Arnold, an ambition that expressed itself in a simultaneously technological and cultural commitment to disseminate over the airwaves what Arnold had called 'the best which has been thought and said in the world'.[1] However, during the Reithian era, the BBC broadcast many programmes that expressed an ideologically more flexible attitude toward broadcasting.[2] During the formative years of the 1920s and 1930s, Hilda Matheson and other innovative producers opened spaces on air for the celebration of modernist literary experimentation and for the promotion of aesthetic and ethical values that ran counter to the company line. As a matter of fact, then, if not of policy, the early BBC encouraged literary-sonic experiments that exemplified a modernist ethos and dramatized the individual and social fragmentation and instability that, for modernist writers of all ideological stripes, defined life in the early twentieth century.

In the steadily growing field of literary radio studies—a field rooted in the methods of literary criticism and groping toward a more vigorous interdisciplinarity—there has been a growing interest in early radio drama both as a marker of the BBC's openness to aesthetic innovation and as a modernist form in its own right. This critical interest is not entirely new: it's been thirty-five years since John Drakakis casually asked that early British radio drama 'be seen in the larger context of experimentation in all the arts during this period'.[3] Since Drakakis made this call, historians and literary critics have refined our understanding of literary modernism's complex relation to the technology of radio and the technocultural form of public service broadcasting (as indeed to the other cultural forms of broadcasting—commercial, totalitarian) at the moment of radio modernism's happening in the 1920s and 1930s.

With respect to radio drama specifically, critics and historians are revealing fascinating connections between it and literary modernism's formal experiments—or, to be more precise, literary modernism's *other* formal experiments. Twenty years ago, Paddy Scannell noticed the presence in early radio drama of ideas and formal elements that derived from the nineteenth-century tradition of aestheticism which exerted a tremendous influence on the development of literary modernism. Scannell observed that when, in 1928, the BBC began exploring the possibilities of radiogenic drama in a 'Special Research' unit, it was 'encouraged by the Victorian notion of art for art's sake'.[4] Scannell offers this remark only in passing, without following through on the implications of such encouragement. I want to take Scannell's remark seriously, and to begin pursuing its uses for the study of radio modernism. The influence of aestheticism on British modernism is a well-trodden field of inquiry in literary studies; it has received, to all intents and purposes, no attention at all in radio studies or media history.[5] And yet, just as aestheticist ideas helped to shape literary modernism, so too did they contribute to the development of early British radio drama. The resemblances between important examples of radio drama and modernist fiction at the turn into the 1930s offers an interesting opportunity for literary critics and media historians alike, to refine our understanding of how, far from representing unconnected examples of cultural production, radio drama and modernist fiction are united both through their shared roots in the art for art's sake movement and through their shared participation in a new mass media ecology. To describe the uncanny formal and thematic similarities between the radio dramas that Tyrone Guthrie produced around 1930 and the radical fictional experiment that Virginia Woolf was pursuing at the same time in her novel *The Waves* (1931) is to deepen our appreciation not only of the cross-pollination between radio drama and modernist fiction, but also of how both of these cultural forms emerged from and reflected the influence of the aestheticist legacy and of a shared technocultural environment permeated by radio waves.

The Aestheticist Background

When I refer to the origins of the aestheticist tradition here, I am thinking first and foremost of Walter Pater's *Studies in the History of the Renaissance* (1873), the book that popularized the provocative idea, cited by Scannell, of 'art for art's sake'.[6] In this collection of essays about Renaissance artists, Pater advocated a non-moral view of the nature and purpose of art that flew in the face of Victorian orthodoxy, articulated by the art critic John Ruskin, for whom good taste was a function of good morals, and Arnold, who hoped that culture might fill the spiritual gap left by the steady withdrawal of religion as a psychologically stabilizing, socially unifying, politically harmonizing force.

Pater inspired a revolution that was at once ethical and aesthetic. The idea that art's value lay in itself and not in its ostensible moral function ramified through *fin-de-siècle* and modernist aesthetics. *The Renaissance* helped to energize the spirit of revolt and innovation that defines modernism; and it consecrated art to an inviolate sacrosanctity—all ultimately in the interest of intellectual, ethical, and aesthetic independence, or what Pater called 'antinomianism', as the *sine qua non* of the good life.[7] For Pater, art for art's sake comprises three core elements: It refers to an ethic; it describes a method; and it expresses an aspiration. As an ethic, art for art's sake is a way of being, a disposition toward the world, an anti-utilitarian celebration of the intrinsic value of art and of the habit of 'burn[ing] with a hard, gem-like flame'.[8] The very claim of art for art's sake releases art from moral servitude and affirms the value of art beyond use.

The method of Paterian art for art's sake is impressionism, which involves the vigilant effort to court and carefully attend to sense impressions that 'flood' the mind from without.[9] Pater's impressionism powerfully influenced literary modernism's understanding of itself; it paved the way for the proto-modernist and properly modernist impressionism of writers like Robert Louis Stevenson, Joseph Conrad, Ford Madox Ford, James Joyce, Dorothy Richardson, Virginia Woolf, and many others. As Max Saunders writes, 'Impressionism was not just the fundamental antecedent to Modernism, but the ground on which modernism was constructed'.[10]

Art for art's sake antinomianism and impressionism influenced the development of modernism as an ethic and a method, respectively. Another aspect of aestheticist theory provides perhaps the most vital link between radio drama and other modernist literature during modernism's heyday. This is the idea that, as Pater puts it, 'All art constantly aspires to the condition of music'.[11] This phrase appears in an essay titled 'The School of Giorgione', which he included in the 1888 edition of *The* Renaissance. In this phrase, Pater's explicit purpose is to affirm art's fundamental freedom from representation and, therefore, its independence from morality. On a different level—and from the perspective not of the producer of music, but of its audience—this aspirational claim bespeaks Pater's attentiveness to the individually transformative power of listening. In 1888 Heinrich Hertz was proving the existence of radio waves, and catalyzing the process of technological innovation that would result in radio as a communications medium. At this same moment, Pater was reflecting on how Giorgione's paintings 'caught the waves of wandering sound'.[12] Indeed, their representations of listeners show audiences eager 'to detect … the smallest undulation in the air.'[13] As a result, in Giorgione's paintings, as Pater puts it in a passage on the erotics of listening, 'life itself is conceived as a sort of listening' when 'the happier powers in things without are permitted free passage, and have their way with us'.[14]

Given his attunenedness to sense-impressions in general, and to sound in particular, if Walter Pater had lived to hear radio he no doubt would have revelled in the reception of its sounds. With his interest in how sensory impressions penetrate 'that thick wall of personality through which no real voice has ever pierced on its way to us',[15] he would have appreciated that to listen to radio is not only to entertain a new type of private sensory impression, but also to participate in a new 'listening public';[16] it is to be seduced into a world larger than the self and to cultivate at once a sense of belonging, of community, and of self-dispersion in a wash of myriad impressions. If Pater had lived to read *The Waves* and to hear the radio dramas of Lance Sieveking and Tyrone Guthrie, he would have recognized spirits kindred with himself and with each other. These writers and artists were grappling with the implications of the scientific discoveries and technological innovations that were reshaping the modern cultural ecology, and arriving at methods of representing the sensory and psychological experiences characteristic of a world shaped by radio.

Aestheticism and Radio Drama: Lance Sieveking and Tyrone Guthrie

In the 1920s and 1930s, at the high-water mark of literary modernism, 'disembodied voices were whizzing about, invisibly, in all directions—over the horizon and across oceans, through solid buildings and into living rooms—all at the speed of light'.[17] At this time the BBC invited writers and producers to participate in, indeed to help shape, the new media ecology created by radio. In some cases, this invitation meant that, despite the BBC's official Arnoldianism, it created an opportunity for writers to become aestheticist-inspired modernist composers of sound. Some took full advantage of this opportunity, as can be seen in the cases of the radio dramatists Lance Sieveking and Tyrone Guthrie. Sieveking and Guthrie conceived their experimental radio dramas at the turn into the 1930s in explicitly aestheticist terms; they were aware of their role as creators of impressions; and they deliberately strove to achieve the condition of music and to encourage new habits of listening—new ways of 'catching wandering sound'. The ways they pursued these aims define Sieveking and Guthrie as Paterians in the radio age.

It is difficult to imagine a more aestheticist statement of purpose than the lyrical passage in which Sieveking recalled how he felt after the completion of his 1928 broadcast *Kaleidoscope I*, a formally and technically adventurous broadcast that represents one of the highlights of the short period, 1928–1932, when, as Sieveking put it, 'Radio began to be talked of not only as a new form of communication, but also as a potential art form'.[18] 'My great experience was over', Sieveking wrote:

> There was nothing left of it. I realized at that moment the terribly fleeting nature of the medium in which I worked. Yes, that's the point. So terribly fleeting.
>
> How solid a structure I had seemed to build. Yet the whole of my complex lovely picture with its voices, its castles, its landscapes, its musics, its men and women, had been painted on the underside of a cloud with a brushful of iridescent vapour. It melted as you looked and listened, and, like so much moonshine, it disappeared for ever.[19]

This passage appears in the opening pages of Sieveking's book *The Stuff of Radio* (1934), at the end of his vivid impressionistic description of the 'absorbing occupation of producing radio-plays'. Such works constitute for Sieveking the true 'stuff of radio'.[20] 'There is only one true stuff of radio', he writes; 'the sound-effect, indeed sound itself, arranged in a certain way, is the essence of radio art'.[21] This statement borders on the tautological: of course sound is of the essence of radio; radio is sound. But in approaching the art of radio drama from this formalist perspective, Sieveking defines the radio dramatist in terms that harmonize with the Paterian observation about art's aspiring to the condition of music. 'The radio dramatist', he says, 'becomes like the composer of music'.[22] Writing of both *Kaleidoscope I* and the following year's experiment, *Kaleidoscope II*, Sieveking describes them as 'rhythms' embodying the lives of a man and a woman, respectively, from cradle to grave. By imagining individual lives as examples of rhythms, and in thinking of his work in radio drama as that of a painter in the fleeting medium of sound, Sieveking was driven by an aestheticist impulse.

Sieveking's aestheticist impulse was shared by his contemporary, the Irish radio dramatist Tyrone Guthrie; both artists, in Paddy Scannell's phrase, were 'encouraged by the Victorian notion of art for art's sake'. Guthrie made a significant contribution to the theory of early radio drama in his introduction to his published collection *Squirrel's Cage and Two Other Microphone Plays* (1931). This introduction also constitutes an aestheticist manifesto of the 'microphone play' in its echoes of Pater's 'Conclusion'. The first several paragraphs of Guthrie's essay comprise a series of generalized reflections on the impressionistic aspects of radiogenic drama. Then, in brief comments on each of the plays, Guthrie reveals a complex aestheticism composed equally of impressionism and a striving toward musicality. In his discussion of *The Flowers Are Not for You to Pick*, a radio play first broadcast in 1930, Guthrie focuses on the play's impressionistic elements, his having 'taken pains … to provide the listener with the material to create his own *décor*'.[23] Discussing the other plays in this collection, *Matrimonial News* (first produced in 1938) and *Squirrel's Cage* (first produced in 1929), Guthrie emphasizes aspects of their composition that point to the influence of Paterian formalism and musicality. In *Matrimonial News*, for instance:

> the effect aimed at is primarily musical. Picture is all the time subordinated to rhythm; the characterization is not subtle, but broad and flamboyant, because the contrast in pitch and colour between one voice and another is more important as a musical contribution to the *ensemble* than as an expression of individual character.[24]

Like *Matrimonial News*, '*Squirrel's Cage* attempts to combine both methods; pictorial in the scenes, musical in the interludes'.[25] And in a passing remark on the plays' preparation for publication, Guthrie explains his idiosyncratic use of punctuation as a sort of musical notation. It 'may seem irregular', he writes: 'It is intended rather to express the rhythm of the lines than their grammatical construction'.[26]

As with Sieveking's *Kaleidoscope* plays, Guthrie's three plays imagine the creator of microphone plays as, in Sieveking's phrase, 'like the composer of music'. *Squirrel's Cage*, for instance, as Howard Fink explains, traces the life of Henry Wilson, a stultified middle-class suburban man, from birth to maturity in a similar kind of 'rhythm' to Sieveking's.[27] *The Flowers Are Not for You to Pick* likewise creates a rhythm of a life, in the story of Edward, a forever belittled young man destined to humiliation and failure in personal and public life. The play's action comprises, as the sound cues indicate, pictures and voices from a drowning man's past. *Matrimonial News* consists entirely of the multivocal interior monologue of Miss Florence Kippings, the frustrated thirty-something daughter of a demeaning shopkeeper mother that occurs entirely 'in the mind. […] Remember you are overhearing her thoughts' as she sits 'alone in a cheap restaurant in the Strand in London'. She is ostensibly waiting for a man whom she has contacted, or imagines she has contacted (it is impossible to know), through the personal advertisements in the magazine that gives the play its title.[28] Like the stream-of-consciousness narration characteristic of so much modernist fiction—the example most appropriate in the present case is the concluding episode of James Joyce's *Ulysses* (1922)—Florence's interior monologue is intended, Guthrie writes, 'to convey the half-formed ideas and images surging, tumbling, churning in the consciousness of Miss Kippings'.[29] It is also intended, Guthrie explains, to be experienced 'more as rhythm and jingle than as sense'.[30]

In his general introduction, Guthrie highlights the intended musicality of *Matrimonial News* and *Squirrel's Cage*. In his brief explanatory note prefacing *The Flowers Are Not for You to Pick*, he explains its own musicality: 'the many short scenes rise out of and sink into a rhythmic sound of splashing, moving seas. This sound should be complex yet symphonic'.[31] Finally, *The Flowers* possesses a rhythm in its alternation of short scenes from young Edward's life—each illustrating a stage of the desperation that leads him to abandon England on a China-bound ship from which he falls into the ocean—with the sound of waves whose own rhythm marks the transitions between scenes. So too in *Squirrel's Cage* scenes from the growing Henry Wilson's life alternate with interludes composed of rhythmic repetitions of voices and of the sounds of machines—trains, typewriters—that reinforce the mechanical routine, the steady, monotonous non-drama, of a life constrained by soul-deadening middle-class suburban professional pragmatism and caution. The play ends with the note: 'Voices cross-fade with Closing Music'.[32]

Lance Sieveking's kaleidoscopes and Tyrone Guthrie's rhythmic repetitions and interior monologues and sound-mosaics import core aspects of the aestheticist tradition —its impressionism and its aspiration to musicality—into a new media ecology created by the advent of radio broadcasting, and a new cultural ecology shaped both by radio and by developments in modernist writing. In doing so, they not only, as Amanda Wrigley shows in *Greece on Air* (2015), helped to establish what would become stock techniques of radio drama over the twentieth century, but also reveal close formal and thematic connections to other modernist experiments in literary form.[33]

Rhythms and Waves: Tyrone Guthrie and Virginia Woolf

Examined together, Tyrone Guthrie's *Squirrel's Cage* and *The Flowers Are Not for You to Pick*, performed in 1929 and 1930, and Virginia Woolf's 1931 novel *The Waves*, which she conceived of as a generically hybrid 'playpoem' and 'wr[ote] to a rhythm and not a plot',[34] represent an interesting case of how radio drama and modernist fiction echo each other's formal methods; they also all articulate creative commentaries on the changing habits of perception in a new electronic mass media ecology. Woolf was a regular radio listener who also broadcast three times at the BBC. Two of these broadcasts happened in 1927 and 1929, or at about the time that Sieveking specifies as the moment when the BBC began to take radio seriously as an art form. Woolf's broadcasts were contemporaneous both with Sieveking's *Kaleidoscope* experiments (in 1928 and 1929) and with the broadcasts of *Squirrel's Cage* (1928) and *The Flowers Are Not for You to Pick* (1930). Because of the close formal and thematic resemblances between Woolf's novel and Guthrie's *Squirrels* and *Flowers*, this case is particularly tantalizing with respect to the question of influence. Did Sieveking's and Guthrie's microphone plays influence Woolf methodologically or thematically, or on the molecular level of phrases and words? No evidence has been discovered that Woolf knew Sieveking's or Guthrie's productions, so it would be going too far to suggest a direct influence. But given Woolf's interest in radio and her alertness to programming developments, it is at least possible to show how her thinking about fiction in the years immediately surrounding the turn into the 1930s, while she was writing

The Waves, represents an expression of an emerging radio mindfulness among modernist writers of a kind that has yet to receive close critical attention.

Several remarks that Woolf made in the years immediately prior to beginning composition of *The Waves* confirm Woolf's fascination with this new communications and artistic medium, and offer glimpses into how she understood radio's potential to transform habits of perception, thought, and feeling—a transformation that finds its way into her novel as structural and thematic elements. In 1927, just as the first pulses began the rhythmic beat in her brain that would become *The Waves*, Woolf revealed a complex understanding of how radio was shaping a new media and cultural ecology and of the technology's existential and affective implications. In a lecture titled 'Poetry, Fiction and the Future' that she delivered at Oxford in May of that year, Woolf testifies to how radio, by permeating the individual consciousness with waves of electronically mediated information, confirms the illusory nature of individuality in the modern world. This is a time when, she writes, despite the increasing atomization of social life, despite locked doors and bolted windows, everyone with access to a wireless set 'is linked to his fellows by wires which pass overhead, by waves of sound which pour through the roof and speak aloud to him of battles and murders and strikes and revolutions all over the world'.[35] Woolf figures this permeation of the private life of the individual in aestheticist terms. For her, this pouring through and linking and speaking aloud constructs the individual as 'the centre and meeting-place of an extraordinary number of perceptions', just as, for Pater, other voices 'penetrate the thick wall of personality' within which we seem to exist as 'solitary prisoners'.[36] Woolf's existential description of moments of perception in the electronically mediated world paraphrases Pater's ethical injunction to 'be present always at the focus where the greatest number of vital forces unite in their purest energy'.[37] In 1927, on the threshold of the first great age of British radio drama, Woolf is already thinking about radio within an aestheticist frame, just as she is beginning to think of fiction in radiogenic terms.

The Waves is composed of interlacing soliloquies spoken by six characters as they grow from childhood to maturity and into late middle-age; these soliloquies are punctuated by lyrical 'interludes' of natural description that trace the diurnal passage of the sun across the sky from morning to night. The final section of the novel is a long soliloquy in which the character Bernard attempts to articulate a philosophy of life as a way of bringing summary order to his fragmented experiences. Following his last words, 'Against you I will fling myself, unvanquished and unyielding, O Death!' a final, brief interlude closes the novel: '*The waves broke on the shore*'.[38] Recent criticism by Pamela Caughie and Jeremy Lakoff reads *The Waves* not only as a novel of the radio age, but also, more pointedly, as a work of radiogenic fiction. Woolf recorded in several diary entries the challenges of writing this novel, detailing over a period of more than four years, 1927–1931, her fascination with her creative process and providing insight into her crafting of a new genre. Woolf imagined, in fact, writing not a novel but a new form of drama. In her diary entry for 21 February 1927, at the same time when she was ruminating on the ideas in 'Poetry, Fiction and the Future', she asked herself, 'Why not invent a new kind of play. [...] Away from facts: free; yet concentrated; prose yet poetry; a novel & a play'.[39] Late the following year, she mused on the possibilities of generic hybridity for her new work: 'That was to be an abstract mystical eyeless book: a playpoem'.[40] Throughout the period whe she was writing *The Waves*, Woolf continued to reflect on the nature of this new genre and on her method for achieving it. So, for example, on 20 August 1930, she records: 'The Waves is I think resolving itself [...] into a series of dramatic soliloquys. The thing is to keep them running homogenously in & out, in the rhythm of the waves'.[41] A diary entry and a letter from a week later note her efforts to construct not a plot but a rhythm.[42]

Because of its attentiveness to how sound structures the identities that Woolf's interconnected characters articulate through speech, *The Waves*, as Caughie writes, 'reads like a radio drama'.[43] Caughie goes on to note that 'sound is not a topic or theme but infuses the very form of the novel'.[44] Moreover, she observes, Woolf's 'sensitivity to aural culture had aesthetic implications'.[45] Caughie figures these implications as explicitly radiophonic:

> The novel is replete with sounds, most obviously the sound of the waves heard throughout the book. [...] But more than any other [of Woolf's works], this novel is *conceptualized* in terms of sound. The six characters speak in soliloquies to an unseen listener [...] using the pure present tense of a broadcast: 'Now [...] the time has come'.[46]

Lakoff extends Caughie's argument in his examination of how individual characters in Woolf's novel occupy 'a soundscape suffused with signals and waves coming from distant origins. [The characters] Louis and Jinny become, in a sense, radio receivers—conduits for the voices and sounds that circulate in the world, reducing them to order'.[47] Lakoff concludes that, although the physical objects of radio sets are absent from Woolf's fiction, the medium of radio itself 'is figuratively present in all its imaginative power'.[48]

A growing body of criticism on radio and modernist literature is showing with increasing nuance how, as Caughie puts it, 'modernist writing and new technologies are [...] reciprocal cultural forms', exerting a 'mutual influence'—both of them participating, in Melba Cuddy-Keane's apt phrase, in 'broad currents of thinking' that shaped social and cultural life in the 1920s and 1930s and permeated the first age of broadcast telecommunications and radio artistry.[49] These currents are audible in Guthrie's plays and Woolf's novel.

A brief announcement sets the scene of *The Flowers Are Not for You to Pick*: 'It is said that their past lives float before the eyes of drowning men. From a ship, bound for China, a young clergyman has fallen overboard ... even now he is struggling for life in the water'.[50] Following this announcement, the play begins with the performance direction: '*The sound of waves*'.[51] Another, elliptical, announcement indicates,'His name is Edward. And before his eyes float pictures ... voices sound in his ears ... voices ... voices ... his past life'.[52] A final opening performance direction states, '*The waves fade as the first scene begins*'.[53] The text of the play then traces Edward's growth from childhood to adulthood, when, to escape a lifetime of humiliating insults to his masculinity and his habit of stuttering, and failures both in his priestly vocation and in love, he embarks on his ill-fated voyage. Like Joycean epiphanies or Woolfian 'moments of being', like the 'eternal moments' of E. M. Forster or the 'symbolic moments' of Joseph Conrad or Thomas Hardy's 'moments of vision', the 'blazing moments' of Katherine Mansfield or the 'decisive moments' of Henri Cartier-Bresson, the play's seventeen brief scenes mark important moments in Edward's life and, as Guthrie explains in his performance directions, '*rise out of and sink into a rhythmic sound of splashing, moving seas*'.[54] In other words, the performance of the play consists in—and Guthrie's listeners heard—the following: a) natural sounds of waves, b) narrative episodes, punctuated by c) interludes of wave sounds, and d) a concluding sound of waves: at the end of the play, Edward drowns; the final performance direction states, '*Waves—then silence*'.[55]

Guthrie's gloss on the existential significance of the '*rhythmic sound of splashing, moving seas*' is fascinating as a summing-up of his ultimate theme of absurdity:

> The sound should be complex yet symphonic [...] by its rhythm and tone it may be possible to suggest not merely the waters in which Edward is engulfed, but the beating of a heart, the tumult of fear, the immutable laws and irresistible strength of Nature compared with our puny and inconstant selves.[56]

Is there a more concise thematic summary of *The Waves*, and in particular of its diurnal interludes ending in darkness and of its final soliloquy, which ends with a defiant stand against death that is undercut by the laconic natural fact stated in the concluding interlude: 'The waves crashed on the shore'? Of course, Guthrie's passage can't have summarized *The Waves*. The question is open, however, whether it, and his scene-interlude structure, may have influenced Woolf. Guthrie's directive appears in the text of the play published in 1931. Woolf would have seen this gloss in the printed script if Guthrie's collection was published early in 1931 and if Woolf read it. The publication date is unclear—whether it appeared before, after, or simultaneously with Woolf's novel on 8 October. And so, even if Woolf hadn't heard the broadcast of *The Flowers*, if Guthrie's play was published at the beginning of 1931, then Woolf would have been able to read it before completing her revisions: she was still revising her novel, and concentrating on the interludes and the final soliloquy, in June 1931.

In the absence of specific evidence it is conjectural to suggest anything greater than coincidence in two facts about Guthrie and Woolf: that *The Flowers Are Not for You to Pick* was first broadcast, on London's Regional Programme, on 10 April 1930; and that, according to a diary entry tallying up 'my Waves account' made on 7 July 1931 and dating the stages of her novel's composition, Woolf completed her first full draft of *The Waves* on the same day when *The Flowers* was broadcast—and then continued to revise it for more than a year.[57] More significant with respect to echoes between *The Flowers* and *The Waves* is the play's, and the playpoem's, emphasis on musicality. The sound of Bernard's soliloquy is complex and symphonic, as Elicia Clements has shown in her examination of Woolf's musical narrative method in *The Waves*. This was a method inspired by Woolf's repeated listening to Beethoven while writing the novel and her new friendship with the composer Ethel Smyth.[58] The 'rhythm and tone' of the soliloquy both suggest the conceptual waters in which Bernard is engulfed. They also describe the tumult of Bernard's fear before the immutable laws and irresistible strength of nature compared with our puny and inconstant selves—selves as ephemeral as the multiple voices of which Bernard is composed. In addition, the rhythm and tone of Bernard's soliloquy also recall the insistently repeating voices that Guthrie's Henry hears while struggling to hold himself afloat in literal waves before they engulf him as *The Flowers Are Not for You to Pick*, anticipating *The Waves*, ends with the sound of waves fading into cosmic silence.

The case for influence here—and further evidence of the cross-pollinating influence of modernist fiction and radio drama—perhaps grows stronger if we bring Guthrie's first microphone play, *Squirrel's Cage*, into the mix. This play is composed of six scenes punctuated by five impressionistic interludes; each interlude features a chorus of voices and rhythmic sounds from the everyday life of Henry Wilson, who desperately wants to escape, as he testifies with breathless urgency, 'the monotony [...] the same old routine day after day—round and round—like a squirrel in a cage' that defines his life, in order to 'feel [...] that we're *progressing, not* just running round and round—and the feeling that all the time one's building up something solid—something sound'[59]—just as, in a novel composed of soliloquies punctuated by impressionistic interlude, Woolf's character Bernard desires 'that something adheres for a moment, has roundness, weight, depth, is completed'.[60]

With its soliloquies and interludes, *The Waves* echoes the dramatic, non-narrative scene-musical interlude structure of Guthrie's play and his concern to represent 'the minute pulsations of feeling, the delicate gradations of thought' that are found in so many works of modernist fiction written under an aestheticist influence. Clements writes that 'interludes' was 'a term Woolf coined for the lyrical passages that describe the developments of a single day'.[61] It is worth noting, though, that Tyrone Guthrie not only developed a similar technique to Woolf's at exactly the same time, but also called his own interludes by that very name and described them as 'musical'.[62] Woolf's 'poemplay' reads, as Caughie puts it, 'like a radio drama', just as, in Howard Fink's words, Guthrie's 'Interludes are pure radio-drama, and the most innovative elements in the play'.[63] The radiogenic qualities of *The Waves* closely resemble those of Guthrie's first two microphone plays, which in turn clearly owe a debt to the stream-of-consciousness methods of narration and psychological exploration pioneered by Woolf and other modernist writers of fiction in the 1920s.

Caughie argues that *The Waves* reads like a radio drama. What this means is that Virginia Woolf writes like a playwright-producer in the mould of Lance Sieveking and Tyrone Guthrie. Like them, she orchestrates a chorus, a symphony, a rhythm. Just as Woolf subordinated plot to rhythm in her 'playpoem' using a new soliloquy technique for dramatizing individuals' essential interrelatedness, so too Sieveking and especially Guthrie dramatized the rhythms of everyday lives through speech aspiring to the condition of music in the increasingly technologized mass culture of the modernist period. If we put aside entirely the question of direct influence, and simply listen closely to the waves of Guthrie and Woolf, what we hear is the extent to which an awareness of how sound waves were reshaping the cultural soundscape permeated literary and radio-dramatic modernism at the turn into the 1930s. Guthrie's plays and Woolf's novel, in both theme and form, embody a fundamental fact of radio modernity: they enact the experience of listening in a new media ecology. They are both, like *Squirrel's Cage*, 'something sound'. *The Waves*, with, in Lakoff's phrase, its 'soundscape suffused with signals and waves coming from distant origins',[64] and *The Flowers Are Not for You to Pick*, with its drowning man awash in memory of 'voices [that] sound in his ears',[65] each, like Walter Pater's Giorgionesque listener ravished by sound, enact the perception-altering, self-disrupting dramatic encounter with powerful waves that threatened to flow through roofs and into homes and to drown the listener in a wash of voices and of mechanical and musical sounds. Both works allegorize the new perceptual experience that Pater himself had felt at the dawn of the age of waves in the 1880s and that was occurring in the minds of listeners everywhere in the 1920s and 1930s. Aestheticism and waves and modernist literature and radio drama are aspects of a complex, shifting, evolving cultural whole—a long cultural moment—that spanned a half-century and found many different permutations.

The new media ecology, compounded of all these elements, also provoked in such modernists as Sieveking, Guthrie and Woolf an ethical response, one also grounded in a shared aestheticist heritage. Like so many modernists, Sieveking follows in the footprints of Walter Pater in celebrating appreciation of the 'moment' as the first step toward enhancing our responsiveness to other faces, voices, persons. Sieveking, in *The Stuff of Radio*, adds a new phrase to the lexicon of modernist moments when he argues that the 'first attraction and interest' of radio drama resides in 'the fact that what is heard is *real*, in that its voices are the voices of men and women alive at that *moment of hearing*'.[66] Sieveking, Guthrie and Woolf discovered new methods in a radio world for creating living characters; in each case, their characters inhabit a world in which they experience others by listening, just as we listeners experience their characters by listening, to broadcast sounds or to printed soliloquies eminently adaptable for broadcast. For Pater, the purpose of the effort to gather sensations—to 'be present always at the focus where the greatest number of forces unite in their purest energy'[67]—is the intensification of the fleeting self we have, for its own sake. The aim of impressionism, as the method of the art for art's sake ethic, is the quickening of the individual consciousness and the enhancement of our attentiveness to the world as, around us, it is—Pater's word—'broadcast'.[68] And broadcast to listeners who ourselves become regenerative circuits, amplifiers of the voices and other sounds that draw us out of our perceptual isolation. In an important contribution to modernist literary radio studies, and to the study of early British radio drama, Angela Frattarola noted in 2009 how some vital experiments in radio drama at the turn into the 1930s are 'reminiscent of Walter Pater's famous pronouncement that the experience of art is dependent on "impressions unstable, flickering, inconsistent, which burn and are extinguished with our consciousness of them"'.[69] The radio dramas of Sieveking and Guthrie, ca. 1930, are more than reminiscent of Paterian impressionism. In Sieveking and Guthrie, art for art's sake ideals, coupled with with the modern fascination with waves, constitute a living presence in early radio drama, just as they do in the modernist fiction of Virginia Woolf. All of these writers were, as Scannell suggests, 'encouraged by the Victorian notion of art for art's sake'. The radio modernism of early microphone plays, and their ethical as well as their aesthetic crux, can be measured by the extent to which they invite us to hear other lives, and to imagine our own, as rhythms approaching the condition of music.

Disclosure Statement

No potential conflict of interest was reported by the author.

Notes

1. Arnold, *Culture and Anarchy*, 5.
2. Avery, *Radio Modernism*, especially Ch. 1, "Arnold over Britain: John Reith and Broadcasting Morality".
3. Drakakis, *British Radio Drama*, 2.
4. Scannell, ""The Stuff of Radio"," 4.
5. In addition to Scannell, see Hendy, "Representing," 34; Hendy, "Painting with Sound"; and Frattarola, "The Modernist "Microphone Play"," 458. Without naming the movement directly, or Pater himself, Hendy's two essays on Lance Sieveking, exploring the aestheticist aspects

of Sieveking's radio dramatic experiments in the late 1920s, represent the fullest examinations to date of art for art's sake influences on the development of early radio drama.

6. Pater, *Renaissance*, 121.
7. Ibid., 16.
8. Ibid., 120.
9. Ibid., 118.
10. Saunders, "Modernism," 429–30.
11. Pater, *Renaissance*, 124.
12. Ibid., 130.
13. Ibid.
14. Ibid.
15. Ibid., 134.
16. Lacey, *Listening Publics*, especially Ch. 2, "The Modernization of Listening".
17. Hendy, *Noise*, 283.
18. Sieveking, *Stuff of Radio*, 15.
19. Ibid., 24.
20. Ibid., 16.
21. Ibid.
22. Ibid., 74.
23. Guthrie, *Squirrel's Cage*, 10.
24. Ibid.
25. Ibid.
26. Ibid., 11.
27. Fink, "Beyond Naturalism," par. 14.
28. Guthrie, *Squirrel's Cage*, 87.
29. Ibid., 86.
30. Ibid.
31. Ibid., 140.
32. Ibid., 84.
33. On early efforts at the BBC to negotiate the boundaries between features and dramas, and on how these efforts helped to generate model forms of dramatization, see especially Ch. 2, "The Contexts of Programme-Making". Also see Williams, *British Writers*, Ch. 4, "Involvements: Writing for the Mass Media".
34. Woolf, *Letters*, Volume 4, 204.
35. Woolf, "Poetry, Fiction," 78.
36. Ibid., 22–23; Pater, *Renaissance*, 119.
37. Pater, *Renaissance*, 119.
38. Woolf, *Waves*, 220.
39. Woolf, *Diary*, Volume 3, 128.
40. Ibid., 203.
41. Ibid., 312.
42. Woolf, *Letters*, Volume 4, 204; Woolf, *Diary*, Volume 3, 316.
43. Caughie, "Virginia Woolf," 345.
44. Ibid.
45. Ibid.

46. Ibid.
47. Lakoff, "Virginia Woolf's," 20.
48. Ibid.
49. Caughie, "Virginia Woolf," 342; Cuddy-Keane, "Virginia Woolf," 73.
50. Guthrie, *Squirrel's Cage*, 141.
51. Ibid.
52. Ibid.
53. Ibid.
54. Ibid, 140.
55. Ibid., 207.
56. Ibid., 140.
57. Woolf, *Diary*, Volume 4, 35.
58. Clements, "Transforming Musical Sounds," 163.
59. Guthrie, *Squirrel's Cage*, 81.
60. Woolf, *Waves*, 176.
61. Clements, "Transforming Musical Sounds," 167.
62. Guthrie, *Squirrel's Cage*, 10.
63. Given its dramatic qualities, it is not surprising that Woolf's novel has been adapted for radio performance several times. Louis MacNeice's adaptation was broadcast in 1955 and 1957, and then again in 1976 and 1977. A dramatization by Terence Davies was broadcast in 2007. For details of the dates and times of broadcast, see: http://genome.ch.bbc.co.uk/search/40/20?q=woolf+the+waves#search
64. Lakoff, "Virginia Woolf's (Absent) Radio," 20.
65. Guthrie, *Squirrel's Cage*, 141.
66. Sieveking, *Stuff of Radio*, 51.
67. Pater, *Renaissance*, 119.
68. Ibid., 118.
69. Frattarola, "The Modernist," 458.

Bibliography

Arnold, Matthew. *Culture and Anarchy*. Edited by Samuel Lipman. New Haven: Yale University Press, 1994.

Caughie, Pamela. "Virginia Woolf: Radio, Gramophone, Broadcasting." In *The Edinburgh Companion to Virginia Woolf and the Arts*, edited by Maggie Humm, 332–347. Edinburgh: Edinburgh University Press, 2010.

Clements, Elicia. "Transforming Musical Sounds into Words: Narrative Method in Virginia Woolf's *The Waves*." *Narrative* 13, no. 2 (May 2005): 160–181.

Cuddy-Keane, Melba. "Virginia Woolf, Sound Technologies, and the New Aurality." In *Virginia Woolf in the Age of Mechanical Reproduction*, edited by Pamela Caughie, 69–96. New York: Garland, 2000.

Drakakis, John. "Introduction." In *British Radio Drama*, edited by John Drakakis, 1–36. Cambridge: Cambridge University Press, 1981.

Fink, Howard. "Beyond Naturalism: Tyrone Guthrie's Radio Theatre and the Stage Production of Shakespeare." *Theatre Research in Canada/Recherches théâtrales au Canada* 2, no. 1 (1981):

n. pag. Accessed February 21, 2017. Web https://journals.lib.unb.ca/index.php/tric/article/view/7519/8578.

Frattarola, Angela. "The Modernist "Microphone Play": Listening in the Dark to the BBC." *Modern Drama* 52, no. 4 (Winter 2009): 449–468.

Guthrie, Tyrone. *Squirrel's Cage and Two Other Microphone Plays*. London: Cobden-Sanderson, 1931.

Hendy, David. *Noise: A Human History of Sound and Listening*. New York: Harper Collins, 2013.

Hendy, David. "Painting with Sound: The Kaleidoscopic World of Lance Sieveking, a British Radio Modernist." *Twentieth Century British History* 24, no. 2 (2013): 169–200.

Hendy, David. "Representing the Fragmented Mind: Reinterpreting a Classic Radio Feature as "Sonic Psychology"." *The Radio Journal: International Studies in Broadcast & Audio Media* 11, no. 1 (2013): 29–45.

Lacey, Kate. *Listening Publics: The Politics and Experience of Listening in the Media Age*. Cambridge: Polity Press, 2013.

Lakoff, Jeremy. "Virginia Woolf's (Absent) Radio." *Virginia Woolf Miscellany* 88 (Fall 2015/Winter 2016): 19–20.

Pater, Walter. *Studies in the History of the Renaissance*. Oxford: Oxford University Press, 2010.

Saunders, Max. "Modernism, Impressionism, and Ford Madox Ford's *The Good Soldier*." *Études Anglaises* Tome 57, no. 4 (2004): 421–437.

Scannell, Paddy. ""The Stuff of Radio": Developments in Radio Features and Documentaries Before the War." In *Documentary and the Mass Media*, edited by J. Corner, 1–26. London: Arnold, 1986.

Sieveking, Lance. *The Stuff of Radio*. London: Cassell, 1934.

Williams, Keith. *British Writers and the Media, 1930–45*. London: Macmillan, 1996.

Woolf, Virginia. *The Diary of Virginia Woolf*. Volume 3. Edited by Anne Olivier Bell. San Diego: Harcourt Brace & Company, 1980.

Woolf, Virginia. *The Diary of Virginia Woolf*. Volume 4. Edited by Anne Olivier Bell. San Diego: Harcourt Brace & Company, 1982.

Woolf, Virginia. *The Letters of Virginia Woolf*. Volume 3. Edited by Nigel Nicolson and Joanne Trautmann. San Diego: Harcourt Brace & Company, 1977.

Woolf, Virginia. *The Letters of Virginia Woolf*. Volume 4. Edited by Nigel Nicolson and Joanne Trautmann. San Diego: Harcourt Brace & Company, 1978.

Woolf, Virginia. "Poetry, Fiction and the Future." In *Selected Essays*, edited by David Bradshaw, 74–84. Oxford: Oxford University Press, 2008.

Woolf, Virginia. *The Waves*. Annotated and with Introduction by Molly Hite. Orlando: Harcourt/Harvest, 2006.

Wrigley, Amanda. *Greece on Air: Engagements with Ancient Greece on BBC Radio, 1920s–1960s*. Oxford: Oxford University Press, 2015.

BBC FEATURES, RADIO VOICES AND THE PROPAGANDA OF WAR 1939–1941

Alex Goody

BBC Radio drama, and especially feature programmes, became a crucial vehicle at the beginning of the Second World War for articulating a sense of national unity and for providing accounts of the war that could counteract Nazi propaganda. This essay will explore the feature series Shadow of the Swastika *(1939–40), Terence Horsley's* Narvik *(1940) and Cecil McGivern's* The Battle of Britain *(1941), examining how the aesthetics of this radiogenic genre developed through encounters with the propaganda politics of the British nation at the outbreak of war and in response to the reactions of radio listeners. For the writers of* Shadow of the Swastika*, the 'sound alone' of radio was crucial for creating an effective actuality in a feature, but the empathetic force of this sound actuality—its ability to stand as/for the real—raised important political and moral questions at a time of national crisis. Examining the broadcasts themselves, and drawing on audience responses to features, the essay will uncover the tensions between political and aesthetic conceptions of the radio feature between 1939 and 1941. As the essay will argue it was in this period, which spans the Phoney War and the first major British military campaigns, that friction arose between anxieties about the presence of the radio voice and its reception by listeners, and the BBC's attempt to use the power of the radio medium to present a real account of the war. Political expediency, technology, radio aesthetics, the exigencies of war and the affectivity of sound were all negotiated through radio features as the Second World War escalated. By 1941, in response to these competing factors, the key parameters of BBC war features had solidified and they would go on to function, in the subsequent years of the war, as white propaganda binding the nation together in the struggle against Nazi Germany.*

In the early years of the Second World War the BBC Department of Features and Drama became a crucial source for articulating a sense of national unity and for providing accounts of the war to counteract Nazi propaganda. Even before the outbreak of war a Defence Subcommittee advised the BBC that 'The maintenance of public morale should be the principal aim of wartime programmes',[1] and the early years of the war saw a range of programming designed to boost morale and foster a sense national unity, from early morning prayers (*Lift Up Your Hearts*) and physical exercises (*Up In The Morning Early*), to *Music While You Work* and variety shows (such as *Helter-Shelter*), to increased educational programming.[2] At the same time the presence of features in the BBC radio schedule expanded. Writers such as Louis MacNeice and Stephen Potter were employed to script and produce features that took on a particular wartime role in broadcasting information and generating a form of official propaganda that could reinforce the shared goals of the Allies.[3] If the monologic power of the fascist radio voice underpinned the success of Hitler's regime in Europe, the features commissioned by the BBC sought to thwart this power with creative expressions of the heterogenic but communal experiences of the British and their allies. However, this essay uncovers the tensions between political and aesthetic conceptions of the radio feature at the beginning of the war. As I go on to argue, it was in the period 1939–1941, which spans the 'Phoney War' and the first British military engagements, that friction arose between anxieties about the presence of the radio voice and its reception by listeners, and the BBC's attempt to use the power of the radio medium to present a real account of the war. Political expediency, technology, radio aesthetics, the exigencies of war and the affectivity of sound were all negotiated through radio features as the Second World War escalated. Although the importance of radio features in the BBC's wartime programming has been well-established in scholarship,[4] this essay critically engages with the paradoxical disembodiments that were navigated in early wartime features and reveals how the deployment of features as official white propaganda was attendant on a curtailment of the affective force of radio voices.

As a 'new' medium, radio had 'emerge[d] into and help[ed] to reconstruct publics and public life',[5] and by the late 1930s the feature, as a nascent radiogenic form, began to play a central role in how BBC radio sought to construct a patriotic public. But the combination of dramatic recreation (dramatised scenes, sound effects, music) with actuality accounts of violence, national struggle and warfare in radio features produced particular, public responses to the influence of radio which impacted the activities of the BBC Features Department at the beginning of the war. As a blind medium the power of radio propaganda lay in its potential to produce affective listening, the intensification of one sense (listening) generating a physical, emotive response to the presence of sound. Siobhàn McHugh explains that 'the affective power of sound and voice, combined with the intimacy of the listening process, means we can be moved by listening' to personal testimonies or accounts.[6] But the hazards of a disembodied aural presence that could engender physical, personal emotions in listeners, generated particular issues as BBC wartime features developed and sought to dramatise contemporary events and actual people. As a result, whilst producing programmes which could bind the nation together in opposition to Nazi Germany in the opening months of the war and as British military campaigns escalated across 1940, the BBC Features Department also had to pay careful attention to what their radio voices might embody for and in their listeners.

While the drama output of the BBC Department of Features and Drama involved adaptations of literary texts for the radio (novels, short stories, plays, Shakespeare) alongside radiogenic work, features were, in the words of Lance Sieveking 'peculiarly, particularly, and integrally the stuff of radio.'[7] Features became increasingly important in the lead up to and early years of the Second World War—Jeanette Ann Thomas points out that BBC feature output grew from only five programmes in 1933 to over two hundred in 1938 -[8] and in May 1936 Laurence Gilliam was appointed Assistant Director in charge of BBC Features, under Val Gielgud who was the BBC's Head of Features and Drama. Features seemed to follow a similar trajectory to the BBC's drama output at the beginning of the war with what Thomas describes as a 'steady stream of "morale" features drawing on the traditions and history of the nation' offering a 'nostalgic portrayal of rural England'.[9] However, with the series *Home Front* that Gilliam proposed in late 1939, images of rural

England were replaced by accounts of contemporary, national personas and regional diversity.

Home Front ran on the Home Service from September 1939 to May 1940 and the first thirty-minute programme in the series, 'Children in Billets' aired on Saturday 30 September (3.30–4 pm). Subsequent programmes included 'The Home Fires Burning' from the Welsh region, 'Harvest of the Sea' and 'Women in War'.[10] BBC staff and freelancers were also commissioned by Gilliam to devise features that provided a background to the news, such as 'The Spirit of Poland', 'The Empire's Answer' and 'All France is Here'.[11] As Gilliam argued in an internal BBC memo in November 1939, feature output 'up-to-date has included a large proportion of what may be called "morale" programmes, dealing with abstract subjects such as patriotism, and democratic virtues', and he suggested 'supplementing the present line with programmes more specifically intended as counter-propaganda'.[12]

The BBC had a 'complex' relationship with the Ministry of Information (MOI) at the beginning of the war,[13] and though it never came under direct government control, the Minister of Information could veto any broadcast and the MOI vetted every broadcaster. For the MOI the BBC had an important role in producing positive propaganda, what Tim Brooks defines as 'White—official—propaganda [...] clearly recognizable for what it was. There was no attempt to hide its origins. It was particularly useful for spreading accurate news about the war or Allied activities'.[14] But the feature, framing and contextualising news and events whilst drawing on the affective force of dramatisation, was a unique propagandist tool. In his memoir *Years of the Locust* Val Gielgud reflected on the wartime 'strength' of 'Feature Programmes' at the BBC, while the writer and producer D. G. Bridson, who described himself as 'joining the BBC when the Feature Programme was coming to birth' highlighted the need for the BBC, in the early months of the war, to 'get radio linked up purposefully behind the national war effort'.[15] Gilliam himself invested heavily in the truth claims of features, emphasising their supposed authenticity in his 'Introduction' to the 1950 volume *BBC Features*:

> In its simplest form, the feature programme aims at combining the authenticity of the [broadcast] talk with the dramatic force of a play, but unlike the play, whose business it is to create dramatic illusion for its own sake, the business of the feature is to convince the listener of the truth of what it is saying, even though it is saying it in dramatic form.[16]

Gilliam's account of the 'truth' of features belies the actual crafted and edited nature of the features produced by the BBC at the beginning of the Second World War. Kate Whitehead describes features as 'something of a hybrid between documentary and drama' that could convey 'the facts in an entertaining, even emotive way' as opposed to the '"straight" reporting of News and Current Affairs Departments', and that took on a 'basically propagandist role' during the war.[17] Sian Nicholas describes features as 'a leap forward in indirect techniques of propaganda' that were consolidated by the BBC in 1942–3,[18] and Gilliam himself reflected that, though features 'fitted no known formula' when they were conceived as a genre in the 1930s, 'the Second World War projected them [features] suddenly to the position of powerful propagandist agencies' because their truths could be simultaneously factual and emotive.[19] But the propaganda power of features was something still being explored by writers and producers at the start of the Second World War. Writing in November 1939, when the potential of features in propaganda broadcasting was still nascent,

Gilliam distinguished between two types of wartime features: firstly he identifies '[p]ositively propagandist' features 'designed to stir directly the national pride of ourselves and our Allies without descending to a jingo level' and includes 'Looke to your Moate' and 'For Ever England' as this type.[20] Secondly Gilliam highlighted features that were '[p]ropagandist by implication only', 'their primary axiom being accuracy of fact and complete balance of point of view'. Gilliam proposed that this model was exemplified by 'The Home Front'[sic] and 'Shadow of the Swastika'[sic][21] a series of features he commissioned in 1939, soon after the outbreak of war.

The Shadow of the Swastika was a series of topical features, based on actual documents, which covered the rise of the Nazis from the First World War to the invasions of Czechoslovakia and Poland. *The Shadow of the Swastika* was broadcast fortnightly from Friday 10 November 1939 (9.25–10.00 pm) to Thursday 29 February 1940 (9.30–10.15 pm).[22] From the outset *The Shadow of the Swastika* was presented to radio audiences in the *Radio Times* as 'authentic' and based on 'documentary evidence' even for the 'fictional conversations' which 'were the things people of that type really did say'.[23] The scripts were written by Igor Vinogradoff and A. L. (Albert Lancaster) Lloyd with music composed by Walter Goehr (under the pseudonym George Walter which Goehr used for his popular pieces). Vinogradoff was a Russian-born historian, Lloyd a staunch Marxist with a particular interest in folk song, and they were both employed as staff writers at the BBC during the war years; Lloyd had experience of wartime features writing on the *Home Front* series and with Gilliam as producer on 'The Empire's Answer'. The eventual schedule for *The Shadow of the Swastika* involved nine episodes, the first and eighth of which were frame pieces, offering an overview of German history and the rise of the Nazis, and which repeated scenes and material from the central six episodes, and the ninth an 'Epilogue' piece based on events in the war so far. The published version of the scripts was organised through the Drama and Features Department by Gilliam and included just six episodes (the broadcast episodes two to seven): *The Rise of a Leader, The Road to Power, The Reichstag Fire, Hitler over Germany, The Shadow Spreads, The Road to War.*[24]

Lloyd and Vinogradoff were not consulted before the publishing agreement was made with John Lane, to Lloyd's obvious annoyance as he had already agreed to serialise the features in the Sunday newspaper *Reynolds News* and was himself looking for a possible book publication. *The Shadow of the Swastika* was directly claimed as the property of the BBC and MOI, who funded the series and distributed hardcopies of the episodes across the commonwealth and to allied nations under the MOI Transcription Scheme,[25] and key editing decisions for publication appear to have been made by Gilliam rather than the authors. The two writers did contribute a Foreword to the volume and were given a £20 bonus on the book's publication. The publication was advertised in *The Listener* thus:

> The script of the series has now been published (John Lane 5s.) in a form specially edited for reading. The nature of, and reasons for, the emendations are explained by the authors in their foreword. The text, however, is essentially that of the original script and should be of equal interest to listeners who wish to recall those very effective broadcasts and to students of the technique of radio drama.[26]

The authors' Foreword, when it refers to the 'emendations' for publication, propose a distinction between 'what sounded good and natural' and what 'looks artificial to the eye', describing the cuts in the scripts of 'indications for effects—like those for music' and 'some entire passages', as necessary to translate the programmes into 'cold print'.[27] *The Listener* announcement, however, elides the aural-print fracture that the Foreword proposes, arguing that the printed page is 'essentially' that of the broadcasts and can function as a prompt to 'recall' the aural experience of the broadcasts and exemplify the 'technique' of the radio feature. That *The Shadow of the Swastika* is here described as a 'radio drama' provides an interesting rationale for some of the changes in the scripts, and also seemingly contradicts the authors' insistence on the factual nature of their radio writing. The elision between *The Shadow of the Swastika* as broadcast features and the *Shadow of the Swastika* [sic] published scripts as drama, highlights how features and drama existed on a fluid spectrum at the BBC.[28]

The status of *The Shadow of the Swastika* as dramatic features with a factual base was reinforced in the opening statement broadcast before each episode:

> Original documents, both published and unpublished, have been consulted. Every quotation and statement of fact have been verified, and the dramatised episodes are all founded on ascertained fact [...] The authors have had the services as consultant of E. L. Woodward, Fellow of All Souls' College Oxford.[29]

Referring to the research foregrounded in this announcement, Siân Nicholas describes the 'unusual care' with which the series was produced and Gilliam's emphasis on 'perfect impartiality' in these features' treatment of recent German history.[30] The emphasis on veracity is reiterated in the published version of the scripts which has a bibliography at the end that lists texts as different as Hitler's *Mein Kampf* (1925), Erika Mann's *School for Barbarians* (1938), on the Nazi's indoctrination of children, and Christopher Isherwood's 1939 novel *Goodbye to Berlin*.[31] Lloyd and Vinogradoff state explicitly in the Foreword that they drew on 'a hundred books', 'newspaper cuttings, confidential reports, and magazine articles [...] in order to work out an authoritative dramatic treatment'; 'We made no statement and dramatized no incident without good authority', they claim.[32] However, as *The Shadow of the Swastika* was funded by the MOI who also provided information for the broadcasts,[33] its 'verified' 'facts' were also directly in the service of anti-Nazi propaganda. Nevertheless the bibliography, with its inclusion of works of fiction (six other novelists besides Isherwood are cited), indicates the writers' interest in creating a credible diegesis for their reconstruction of twentieth-century German history, alongside the generation of authentic radio voices which could channel official white propaganda.

The writers' Foreword to *The Shadow of the Swastika* acknowledges the propaganda aspect of the broadcasts but claims this is 'nothing new to the BBC' while emphasising the 'facts' of these features.[34] An implicit distinction between the verified facts of *The Shadow of the Swastika* and Hitler's propaganda machine and his rise to power was also made in the second broadcast episode. Hitler's first speech in *The Rise of a Leader* is concerned foremost with propaganda:

> You must get publicity; else you will never become known, Propaganda! That's all that counts! If this party is to prepare the way for the national resurgence, we've got to have a programme! Without a programme we can't get at the big public. And that we've got to do. Gentlemen, [fading] listen to me![35]

This initial speech by Hitler, which associates him primarily with publicity, propaganda and a programme, emphasises the purpose of *The Shadow of a Swastika* features series in exposing and countermanding his propaganda. Hitler's exhortation 'Gentlemen, listen to me!' is a telling detail which foregrounds the aural space as one that Hitler's fascist programme sought to dominate. Other episodes reflect on the propaganda machine that underpins Hitler's regime. *The Road to Power* highlights the importance of the German businessman and politician Alfred Hugenberg's media empire in the Nazi's rise to dominance in German nationalist politics, while in *The Road to War* Hitler is presented 'speaking from Nuremberg' 'on the wireless', and his speech at the *Sportspalast* in Berlin (4 September 1940) is introduced with the detail 'Hitler stands on the flag-draped platform and shouts into the microphone'.[36] *The Shadow of the Swastika* pursues an alternate to the Nazi 'programme', opening up the radio broadcast and the public it can reach to voices which counter the logic of Nazi propaganda. As such it appears both formally and thematically aware of the role radio played in making 'imagined communities more tangible' and reformulating 'intimate publics' into a 'ritual of national identity' which could have a crucial political impact.[37]

Against the monologic of Hitler's propaganda, the radiogenic *The Shadow of the Swastika* combines noise and music with speech to create an aural space that exceeds the control of a singular radio voice. The broadcasts thus make good use of a multiplicity of radio speech, particularly various forms of technologically mediated voices; loudspeakers, radio broadcasts, news announcers, telephone conversations. The technological reproduction of voices are foregrounded in particular ways and serve to delineate the distance between the source and sites of reception of speech. Telephone conversations, for example, illustrate the contraction of time and space through communication technology, but also its failure, such as the Austrian Chancellor's unsuccessful attempts to reach Mussolini on the telephone in *The Shadow Spreads*.[38] Telegrams are dictated and read aloud, exemplifying the translation of speech into recorded language that the episodes themselves embody in the transformation of real events into a written script which is then realised as broadcast sound. Amplified voices, marked as 'Loudspeaker' voices in the PasB (programme-as-broadcast) scripts to distinguish them, are presented as the mouthpiece of 'the enormous propaganda machine of Dr. Göbbels' [*sic*].[39] At points, an explicit tension is staged between radio voices, such as in *The Reichstag Fire* where the Narrator's factual account of the German elections in March 1933 (which was immediately succeeded by the Enabling Act which gave Hitler dictatorial powers) is intercut with a 'Radio Voice' reading regional poll results, the sound of an 'enormous cheering crowd—"Sieg Heil, Sieg Heil"', and conversations in the street of worried citizens and Nazi supporters. The scene ends with the Narrator moving from these sounds and voices of disruption, violence and oppression to symbols that encapsulate the Nazi totalitarian state; 'The card index, the rubber truncheon, and the concentration camp are symbols of the new order'.[40] The move from a sonic scape to figurative language in the script signals the triumph of Nazi rhetoric in which fascist ideas become a physical presence in the world.

As with the *The Reichstag Fire* scene, the broadcasts also use acousmatic sounds recorded in Europe, of marching feet, machine gun fire and cheering crowds, as a central part of their aural-scape. In an article for the *Picture Post* in March 1940, Gilliam emphasised the factual nature of these sounds:

> Hundreds of records in the B.B.C. Record Library were examined for authentic sound effects. The cheers of the Nazi throngs at the Nuremberg Congress, at the German entry into Vienna, at the Berlin Sportspalast, in the Reichstag, were extracted from recordings made at the time and collected for use in the production so the atmosphere should be as authentic as the text.[41]

This claim to an 'authentic' noise proposes another level of veracity (that of actual sound as opposed to sound effects) which also relies on technological mediation (in this case recording). The issue of sound effects concerned radio sound engineers and producers in the early years of radio drama and, as Neil Verma discusses, by the late 1930s broadcast sound effects were becoming standardised and, in effect, decoupled sound from its supposed object-source.[42] Gilliam's representation of the broadcast sound effects for *The Shadow of the Swastika*, however, counteracts any sense of aural alienation that may result from the presence of recorded sound in a live broadcast. Indeed the veracity of the 'text' is reinforced by the technologically mediated historical source, producing an authentic aural atmosphere that is not reliant on live presence alone.

To conclude the published scripts of *Shadow of the Swastika* [sic] Lloyd and Vinogradoff chose a significant ending from the broadcast episodes that brings together sound, voice and truth:

[*Martial music, and after a few seconds the voices come in, measured and pompous to the beat of the music.*]

1st NAZI: German people, your country is at war!

2nd NAZI: The war will make Germany greater and richer!

1st NAZI: German men, your country needs you!

2nd NAZI: The war will make Germany greater and richer!

1st NAZI: You were born to die for Germany!

2nd NAZI: The war will make Germany greater and richer!

TOGETHER: You were born to *die* for Germany!

[*Music*]:

[pause]:

WOMAN: Men and women of Germany! *Were* you born not to live? Were you born to die for —Hitlerism?[43]

This conclusion juxtaposes the rhetoric of Nazi propaganda—amplified by the aural presentation of the statements (martial music, pompous intonation, repetition, volume and male voices)—with the quiet commonsense questioning offered by a woman's voice. The female voice is deployed as a gendered counter to the masculinist rhetoric of the Nazis, signifying a maternal connection to nurturing, birth and life.[44] The 'Woman' opposes the empty rhetoric of the German voices with tone, content and the absent body that her gendered voice evokes, exemplifying the effective staging through sound that characterises the success of *The Shadow of the Swastika*. However, this was actually the end of the eighth episode broadcast on February 8, 1940 (a framing overview episode called *From War to War*). *The Road to War* (which was the last script included in

Lloyd and Vinogradoff's published *Shadow of the Swastika* [sic]) ended in its broadcast version on January 25, 1940 with recordings:

(CHAMBERLAIN: 1ST RECORD DOWN TO 'STATE OF WAR EXISTS BETWEEN US')
NARRATOR: The Prime Minister ended his speech with these words:
(CHAMBERLAIN: 2ND RECORD ENDING WITH 'THE RIGHT WILL PREVAIL')
NARRATOR: They are evil things that we shall be fighting against.
(RECORD: HITLER, ENDING WITH CHEERS)
(ORCHESTRA, HITLER THEME TO FINISH)[45]

The sounds that bring the seventh episode of *The Shadow of the Swastika* to a close at the moment of the outbreak of the Second World War are the actual voices of the British and German leaders, examples of Gilliam's 'authentic' aural atmosphere. Using recordings in this way, with their tangible connection to the physical speech of Chamberlain and Hitler, this conclusion moves *The Road to War* securely into the space of actuality and away from the emotive and dramatic reconstructions of the preceding broadcasts that self-reflexively deploy the technology of broadcast.

The Shadow of the Swastika broadcasts were an important success amongst contemporary critics and listeners: 'Here at last is a programme of moment' commented Grace Wyndham Goldie in *The Listener*, going on to note 'the series presents to us and to the world a clear indictment of the Nazi party [...] to the best of my knowledge it is the first deliberate propaganda we have ever had in dramatic form from broadcasting.'[46] *The Listener* on July 25 1940 reported that 'The BBC Listener Research Department calculated that about twelve million people listened to "Shadow of the Swastika"' [sic],[47] the largest ever audiences for feature programmes. Similarly, a Listener Research Memo from Robert Silvey (BBC Listener Research Director) to Gilliam concluded that, overall, *The Shadow of the Swastika* made 'a great impression on listeners' and that 'it has clearly been highly successful.'[48] The programmes were broadcast internationally and claimed a significant mention in Gilliam's August 1940 article, 'The Job of the Radio Feature Man Today', for *London Calling*, the monthly magazine for the BBC's Overseas Service shortwave broadcasting service.[49] Here Gilliam described features such as *The Shadow of the Swastika* as 'treating in dramatic form the events and personalities of contemporary history' and offers a striking analogy for the series: 'No gangster serial has ever had so large or so enthralled an audience as these programmes [...] It was the finest gangster story ever. And it was true.'[50] The tensions between the 'gangster' popularity of *The Shadow of the Swastika* and its truth-claims are manifest here, with Gilliam evoking the drama and suspense of a popular genre which enthrals audiences whilst asserting the veracity of the radio content. Silvey's listener research on the series pointed to its 'great impression on listeners' and its success in 'catching hold of the public's imagination'; he observed that 'Listeners' reactions to it seem to be affected to an unusual extent by subjective considerations'.[51] This account of the response to *The Shadow of the Swastika* ('impression', 'imagination', 'affected' and 'subjective') evokes just that sense of affective listening that the drama of personality, recent events and sound could educe in the feature form. But there is a sense of anxiety about *The Shadow of the Swastika* and its relationship to truth, personality and the impression on listeners which lies behind, for example, the decision to withhold the name of the actor playing Hitler until after the final broadcast.[52] In his *Picture Post* article, published in March 1940, Gilliam revealed that Marius Goring had been chosen to play Hitler after a lengthy search which included rejecting any amateurs 'who had achieved enormous success in their own circle by a dexterous use of the small black moustache, the down-sweeping lock of hair, and the up-lifted arm', specifically refusing the idea of a Hitler impersonator.[53] Nevertheless he describes the two-week audition stage, which involved extensive exposure to different performances of Hitler's voice, as 'a nightmare memory' highlighting the perils of proximity to a vocal personation of the Nazi dictator.[54]

The *Picture Post* article on *The Shadow of a Swastika* also provides a visual record of Goring's radio role in devoting the entire front cover to an illustration of him at the microphone, subtitled 'The Voice of Hitler'. Inside, a full-page photograph of Goring in a recording is paralleled to a small photograph of Hitler, with the body postures and hand gestures of the two men in the images an exact mirror of each other. The photograph caption declares:

> Marius Goring as the Führer. He has heard Hitler speak many times over, He himself speaks perfect German. He has listened to records of the Führer until he knows every pitch and intonation of his voice. Even his gestures—though they are never seen—unconsciously imitate those of Hitler. When he goes before the microphone he is a man possessed.[55]

The photograph and accompanying words work hard to bring to visibility Goring's microphone performance, mentioning the 'never seen' similarities to Hitler's 'gestures'. The uncanny suggestion that Goring is 'possessed' contrasts to Gilliam's more measured account of Goring-as-Hitler which attempts to return his performance to a disembodied realm of sound. Gilliam wrote that Goring, who had lived in Germany, has 'a psychological insight … [into] the mental and vocal image of the Fuehrer', suggesting that, despite the accuracy and impact of Goring's performance, it remains a safely incorporeal one.[56]

The impact of radio voice and sound was something at the forefront for the writers of *The Shadow of the Swastika*. In their Foreword Lloyd and Vinogradoff describe how

> It is impossible for the printed page to give the same impression of the tension of such scenes as, let us say, that in which the Jewish doctor is operating on the child who has been drilled in anti-Semitism at school and in the Hitler *Jungvolk*. Over the air, the contrast between the child's clear dreamy voice and the brutal poem he is reciting, between the exquisite ghostly music and the matter-of-fact efficiency of the doctor's voice, created a dramatic effect which could never have come off in film or theatre, for the whole point lay in sound and sound alone.[57]

Lloyd and Vinogradoff refer here to a scene in *Hitler Over Germany* (broadcast on December 28 1939) in which Dr Aarons, played by Bryan Powle, one of the BBC's Repertory Company actors, operates on the young boy Karl who has been accidently shot during pistol practice. The presentation of the indoctrination of children here, which underpins this scene, draws on Erika Mann's *School for Barbarians* and culminates in Karl's parents' realisation that they are 'Afraid of our own children', a terrifying possibility that George Orwell would later deploy in his depiction of the Parsons' children in the totalitarian dystopia of *1984* (published in 1949). As Lloyd and Vinogradoff describe, the surgery scene relies both on the explicit contrast between Aarons as the '[*calm*]' and '[*quiet*]' doctor deploying a scalpel to save his young patient and Karl's unconscious outbursts ('*Sharpen your long knife in the alley*', 'Kill the Jews! Kill the Jews!'), and on the aural space which contrasts a child's voice and a doctor's patience over a musical score that is quietly eerie.[58] The idea of 'sound alone', voice and music isolated from visual clues is important to Lloyd and Vinogradoff's conception of their feature, and highlights the central dynamics of its success. The 'sound alone' functions to create an actuality but also an empathetic force which, through the very disembodiments of radio, evokes an affective response in the listener.

That *The Shadow of the Swastika* works as 'sound alone' is also emphasised in the Drama and Feature Department's response to interest in the script from film producers. Requests from Twentieth-Century Fox and Pyramid Pictures to buy the film rights in early 1940 were met with concern about the aesthetic and political implications of translating the features into a visual medium. Advice was sought from the MOI, and the Deputy Director of the Film Section of the MOI, G. Forbes, offered a detailed response after reading the scripts: 'The scripts are brilliantly written to produce the desired effect in the chosen medium, i.e. sound, without visual images', he observes, but '[t]o attempt to follow them closely in a film by simply adding visual images would be artistically inept and foredoomed to failure'. Forbes also points to the 'practical difficulty' of the lack of '"library" picture-material [...] for more than a minute fraction of the scenes to be portrayed' and concludes 'that to make staged shots of the scenes would be costly and would entail representation of living statesmen etc., to which the Foreign Office would not consent in any case.'[59] That 'sound alone'—in the form of the radio feature—is an acceptable presentation of factual history, whereas the visual 'representation' of 'living statesmen' would be denied consent by the Foreign Office, exemplifies how the supposed incorporeality of radio voicing meant it could avoid the kind of censorship that beset wartime visual media. The BBC's lawyer R. Jardine Brown agreed with the MOI advice and the, quite lucrative, offers from Twentieth-Century Fox and Pyramid Pictures were turned down.

A few months after *The Shadow of the Swastika*, another feature was aired which dealt directly with the presentation of recent history and caused a controversy around the representation of living (and dead) individuals on the radio. *Narvik* by Terence Horsley was broadcast on Whit Monday 13 May 1940 (9.35–10.15 pm) and it offered, as the *Radio Times* described 'A dramatic reconstruction of a great naval feat of arms' including the 'well known' 'Admiralty message to the Hardy's captain—"Use your own judgment"'.[60] The *Radio Times* claims that 'Terence Horsley, with the help of Admiralty officials, has written this dramatic reconstruction of Narvik' and that '[y]ou will hear the sort of things men said to one another, the way they thought, and how they actually went into action against a greatly superior force.'[61]

The feature was commissioned at short notice from Horsley, a journalist, pilot with the Fleet Air Arm and later editor of *Empire News*, probably as he was simultaneously co-writing *Norway Invaded: The First Full Story*, with James Tevnan. Published in May 1940 as part of the *Cherry Tree War Special* series by Withy Grove Press, *Norway Invaded* covered the combat in the seas around Norway and its invasion by the Germans in April; the fourth chapter was devoted to 'The First Battle of Narvik' on 9–10 April 1940.[62] The script of *Narvik* closely follows Horsley and Tevnan's book, drawing directly on the Admiralty communiqué on the battle, issued on 23 April 1940 and the eye-witness account of Petty-Officer Neal of the *Hardy* both of which are reproduced verbatim in *Norway Invaded*. As a feature, therefore, although it was much less extensively researched than Lloyd and Vinogradoff's *The Shadow of the Swastika*, *Narvik* was closely based on the factual reports included in Horsley and Tevnan's *Norway Invaded*; for its details, including minor events such as the encounter with a Norwegian soldier at the beginning of the feature and large portions of the dialogue, Horsley's *Narvik* draws heavily on Petty-Officer Neal's account.

Narvik endeavours to create a sound drama from the eyewitness testimony, with many combat sound effects and directions for pace, indicating in the script, for example, that for a key action scene the actors should 'Play the following fast'.[63] At another point Horsley's script notes of the Narrator's speech that 'This must be given as a running commentary, like a football match',[64] a rather awkward analogy for the actor reading, given the intense combat scene he is narrating. Reviewing *Narvik* in *The Listener*, Goldie follows the *Radio Times* in describing it as 'a "dramatic reconstruction" of the naval action' and evaluates its 'attempt to re-present for us over the microphone the living drama of our violent time' as 'unusually successful'. She also highlights the tensions between actuality and drama that beset the early wartime features of the BBC:

> It immediately raises the whole point of what productions like this are trying to do. Are they merely an extension of the news service? Do they, in fact, correspond to the quick books of journalism? Are they only endeavouring to give us fresh details about events which we have been following with breathless attention? [...] For they do not confine themselves to giving us new and detailed facts. With their music, their sounds and their dramatic speech they endeavour to create in the listener a dramatic emotion. And it is just because of this that programme like *Narvik* are so full of dangers and so full of opportunities.[65]

Horsley's *Narvik* was drawn from his 'quick book of journalism', but adapted into a radio broadcast it moves into an aural space of 'dramatic' sound and speech and, as Goldie suggests, it attains an affective force capable of producing 'dramatic emotion' in the listener. But, alongside these 'opportunities', the 'dangers' Goldie identified in a feature like *Narvik* were realised in one particular listener's response to the BBC broadcast.

There were many British casualties in the Battle of Narvik, including the Captain of the *Hardy*, Bernard Warburton-Lee, who is played as the (unnamed) 'Captain' in the feature. Whilst another listener declared, after the broadcast, that the BBC should 'not permit vulgarly sensational attempts at the dramatization' of the 'glorious acts' of 'Navy, Army, and Air Force',[66] Mrs Elizabeth Warburton-Lee, the captain's wife, wrote to *The Times* with a particular protest:

> Had they told the story simply and correctly it contained all the drama necessary and no one would have listened to it with greater pride than I. But to impersonate the voices of the living and the dead is unpardonable.[67]

Val Gielgud offered his apologies for the 'deep distress' caused, in a response published in *The Times* a day later, countering criticism by claiming the script for *Narvik* had been submitted for scrutiny to the Press Division of the Admiralty and to a surviving crew member.[68] But Archibald Southby, MP for Epsom, raised the central issue of his constituent, Elizabeth

Warburton-Lee's, complaint in the House of Commons on 29 May, condemning the 'impersonations of both living and dead persons [...] over the air', and asking the Minister of Information, Duff Copper, to 'prevent the recurrence of similar action in the future'.[69] When *Narvik* was repeated on the Empire Service on 28 May this compounded the issue further and on 12 June Southby discussed the issue again in the House of Commons, describing the 'unpardonable bad taste' and 'revolting innovation' of personating serving troops.[70] In response, Cooper informed the House that the BBC deeply regretted any pain caused by the broadcast and assured that 'he had satisfied himself that machinery now existed whereby he could exercise complete control over what was said upon all important political matters'.[71] The radiogenic feature form may have been evolving in response to aesthetic and technological developments and playing a central role in the white propaganda of the BBC during the opening months of the war, but as Britain moved into active conflict in Europe the reconstruction of great exploits had to be referred to a set of procedures distinct from these aesthetic, technical and propaganda concerns. This arises not simply because of the danger of representing 'real' events, but because of the affective force of radio impersonation.

That BBC feature policy changed after the *Narvik* controversy is apparent. In Gilliam's August 1940 article, 'The Job of the Radio Feature Man Today', he distances feature output from the reconstruction of battles and the personation of active servicemen, noting that 'in the war itself, in the operations of the fighting Forces, radio has found grand material for feature treatment' but makes it clear that this material is 'Not in the battle-fronts' which, he states 'is the sphere of the news reporters.'[72] Despite Gilliam's understanding of the propaganda potential of features the *Narvik* controversy impacted on the BBC Feature section's conception of its relationship to current events, and in February 1941 there were still restrictions in place on features with Gilliam conjecturing that '[t]here is considerable hope of removing the ban imposed on us last year in dealing with war-time exploits of the Services, provided we can avoid direct impersonation of identifiable members of the Services alive or dead.'[73] When BBC Features did turn again to directly covering the battle-front exploits of the Services on 8 May 1941, it was to follow carefully the lead of a government ministry (the Air Ministry) and produce a feature that was, as far as possible, an exact reconstruction of government propaganda.

The hour-long feature *The Battle of Britain*, written and produced by Cecil McGivern, was based directly on the Air Ministry pamphlet *The Battle of Britain: An Air Ministry Account of the Great Days from 8th August–31st October 1940*, authored by Hilary Aiden St. George Saunders and published two months earlier in March 1941.[74] Saunders wrote detective and spy thrillers, and mystery novels in collaboration with John Leslie Palmer under the pseudonym Francis Beeding. His propaganda pamphlet for the Air Ministry, written 'chronologically in a clear, racy and accessible narrative style',[75] became an international bestseller, with 4.8 million copies sold in Britain by end of 1941 and 42 editions (including an illustrated one for children) appearing in 24 languages by 1942.[76] McGivern's *The Battle of Britain* was similarly popular; the news editor of the *Daily Mail* wrote to the BBC asking, as a result of 'enquiries from several of our readers', for a repeat of the programme which had been 'most enthusiastically received' after its first broadcast on the Home Service on Thursday 8 May, 1941 (8–9 pm).[77] *The Battle of Britain* was broadcast again on Friday 8 August, 1941 on the Home Service (9.35–10.35 pm) with a revised script and new production, and this production was rebroadcast on Monday 14 September, 1942 (Home Service 9.25–10.25 pm) with a further new production on Wednesday 22 September, 1943 (Home Service 9.40–10.40 pm). The feature was published in a collection of McGivern's features, *Bomb Doors Open and other Radio War Features* in 1941, with this published version of *The Battle of Britain* closely based on the revised PasB script for the 8 August 1941 broadcast. The key change to the initial script was the introduction of a German and a 'Rome Woman' announcer reading misinformation bulletins on the air raids on London; McGivern's published script includes this addition which highlights the difference between the propaganda circulated by the axis powers and the facts of the Battle of Britain as presented in the feature. Writing in *The Listener*, Goldie acknowledged the success of the feature, describing it as 'the type of propaganda which the BBC is right to give us'.[78]

Though using available dramatic and acoustic devices, *The Battle of Britain* achieved success and avoided controversy by not naming any characters, by including only actual recordings of the voice of Hitler, and by following very closely the style and content of the Air Ministry pamphlet. Nicholas argues that the 'painstaking verisimilitude' of features such as *the Battle of Britain* meant such programmes had 'a new and effective appeal to the listener, achieving the impact of drama [...] while avoiding the personalisation that had so pained critics of *Narvik*'.[79] By deferring to official white propaganda and suspending the powerful presence of the individualised radio impersonation, *The Battle of Britain* and subsequent wartime features did indeed avert the criticisms levelled at *Narvik*. McGivern's feature foregrounds a government version of battle-front events and cautiously manages the intimate power of radio voices: it was dramatically 'effective' in its goal as propaganda, as Nicholas argues, but not affective or subjective in its impression on listeners' imagination, thereby avoiding the charged interpersonal connection of listening to an actual person voiced on the radio.

The Battle of Britain and the responses to BBC features explored in this article highlight how the sounds and voices of radio had to be carefully negotiated in the first eighteen months of the Second World War. The famous MOI campaign exemplified in 'careless talk costs lives' propaganda images highlighted the perils of what was 'said' in wartime, but what was 'said' by voices in BBC features, impersonating real individuals, offered both particular opportunities and particular dangers. In curtailing the force of the personalised voiced testimony on radio and focusing on an objective verisimilitude and officially sanctioned accounts of events, features secured a central place in the white propaganda efforts of the BBC.[80] However, what was sacrificed to this war effort was some of the most innovative, individual and self-reflexive use of the 'sound alone' of radio that had been crucial to the success of ground-breaking radio features like *The Shadow of the Swastika*.

Disclosure Statement

No potential conflict of interest was reported by the author.

Notes

1. Report of the Defence Subcommitte on War Time Programmes, January 11, 1939; quoted in Briggs, *The History of Broadcasting in the United Kingdom, vol. 3*, 87–8.

2. On the wartime role of the BBC see Briggs, *The History of Broadcasting in the United Kingdom, vol. 3*; Nicholas, *The Echo of War*; Havers, *Here is the News*.
3. On Louis MacNeice's wartime features see Wrigley, "Introduction," 9–17.
4. On the development of BBC Features and their propagandist role see Whitehead, *The Third Programme*, chapter 6 'Features'. Critical accounts of the role of features in the BBC's wartime programming are given in Thomas, *A History of the BBC Features Department 1924–1964* and Nicholas, *The Echo of War*. Amanda Wrigley points out that '"propagandist" and "morale-boosting" seem to have been used a synonyms by the BBC at this time; [features] were designed to actively support the Allied cause without the broadcast of outright lies', Wrigley, "A Wartime Radio *Odyssey*," 82, *n*3.
5. Gitelman, *Always Already New*, 26.
6. McHugh, "The Affective Power of Sound," 496. McHugh is specifically focused on the affective impact of oral history on radio, but her framework is also relevant for considering the power of voice and sound in features that use (dramatised) individual voices to articulate contemporary historical events.
7. Sieveking, *The Stuff of Radio*, 25–6.
8. Thomas, *A History of the BBC Features Department 1924–1964*, 68.
9. Thomas, *A History of the BBC Features Department 1924–1964*, 110.
10. 'Children in Billets', aired on Saturday 30 September (3.30–4 pm) and was written by A L. Lloyd (a BBC staff writer) and Stephen Potter; 'The Home Fires Burning' aired on Wednesday 18 October (8.30–9 pm) and was written and produced by T. Rowland Hughes (who led on features for the Welsh region); 'Harvest of the Sea' aired on Monday 23 October (8–8.30 pm) and was devised and produced by Alan Melville (a BBC staff-writer); 'Women in War' aired on Sunday 19 November (7–7.30 pm), and was produced by Olive Shapley (a freelancer for the BBC who produced *Woman's Hour* between 1949 and 1953).
11. 'The Spirit of Poland' aired on Tuesday 3 October, 1939, 8–8.30 pm (author/producer unknown); 'The Empire's Answer' by Gilliam and A.L. Lloyd aired on Friday 6 October, 1939 8.30–9 pm; 'All France is Here' by Moray McLaren the assistant director of BBC Features aired on Saturday 4 November, 1939, 10.30–11.15 pm.
12. 'Draft Statement on Policy of Features and Drama in Time of War. For Home Service Board Friday 1st December 1939', 21 November 1939', BBC WAC R19/352/1. Gilliam reports on all the features mentioned above in this memo.
13. Nicholas, *The Echo of War*, 16; for detail on the BBC and the MOI see Briggs section IV chapter 2 'Controllers and Controlled' and Nicholas pp.16ff.
14. Brooks, *British Propaganda to France*, xvii.
15. Gielgud, *Years of the Locust*, 181; Bridson, *Prospero and Ariel*, 14, 77.
16. Gilliam, "Introduction," 10.
17. Whitehead, *The Third Programme*, 110, 111.
18. Nicholas, *The Echo of War*, 65.
19. Gilliam, "Introduction," 9.
20. Looke to your Moate' was broadcast on Sunday 21 March, 1937, 9.05–9.50 pm; written by the historian Arthur Bryant it was produced by Gilliam. 'For Ever England' was broadcast Thursday February 1, 1940, 7.30–9.10 pm; compiled by Val Gielgud Head of BBC Drama it was produced by Moray McLaren.

21. Gilliam, "Draft Statement on Policy of Features and Drama in Time of War. For Home Service Board Friday 1st December 1939," 21 November 1939', BBC WAC R19/352/1.
22. Nicholas notes that *The Shadow of the Swastika* 'originated in requests from a number of organisations, including the Board of Deputies of British Jews, for a popular history of the Nazi movement that might counter press allegations that German anti-Jewish propaganda had some justification' (*The Echo of War*, 150–51).
23. 'Expert on Nazis', *Radio Times* 840, November 3, 1939, 3. This edition of the *Radio Times* also included a one-page extract of the script from the first episode, accompanied by archive photographs of Hitler in 1932 and 1933 (p.8).
24. The broadcasts were as follows: *The Story of the Nazi Party*, Friday November 10, 1939, 9.15–10 pm; *The Rise of a Leader*, Friday November 17, 1939, 9.45–10.30 pm; *The Road to Power*, Wednesday November 29, 1939, 10.15–11 pm, *The Reichstag Fire*, Thursday December 14, 1939, 8.15–9 pm; *Hitler over Germany* Thursday December 28, 1939, 7.45–8.30 pm, *The Shadow Spreads* Thursday January 11, 1940, 6.45–7.30 pm; *The Road to War* Thursday January 25, 1940, 7.45–8.30 pm; *From War to War,* Thursday February 8, 1940, 7.45–8.30 pm; *The Nazis at War* Thursday February 29, 1940, 9.30–10.15 pm. These first broadcasts were on the Home Service and recordings are extant in the BBC Sound Archive at the British Library, London: these are erroneously catalogued under the title *Under the Shadow of the Swastika*.
25. Records of the MOI transcriptions and distribution of *Shadow of the Swastika* can be seen in the file *Transcriptions: Shadow of the Swastika*, BBC WAC, E17/192.
26. *The Listener*, 60, July 25, 1940, 125.
27. Lloyd and Vinogradoff, *Shadow of the Swastika*, 12, 11, 12, 10.
28. The lack of a firm distinction between some features and the radio play format is discussed in Drakakis, "Introduction," 8–9 and Whitehead, *The Third Programme*, 109–111.
29. 'Opening Announcement at 10.15 pm: BBC Home Service Wednesday 29th November, 1939, 10.15–11.00 pm' (*The Road to Power*) BBC WAC 19/1142/1.
30. Nicholas, *The Echo of War*, 151.
31. Lloyd and Vinogradoff, *Shadow of the Swastika*, 194.
32. Lloyd and Vinogradoff, *Shadow of the Swastika*, 'Foreword', 9.
33. This relationship with the MOI is clear from the letters contained in BBC WAC 19/1142/1; see also Nicholas, *The Echo of War*, 151.
34. Lloyd and Vinogradoff, *Shadow of the Swastika*, 'Foreword', 10.
35. (*The Rise of a Leader) The Shadow of the Swastika 3*, November 17, 1939, 7, Play Library Scripts BBC WAC.
36. (*The Road to War) The Shadow of the Swastika 7*, January 25, 1940, 7, Play Library Scripts BBC WAC.
37. Douglas, *Listening In*, 23–4; Loviglo, *Radio's Intimate Public*, xv, see also Ehrick, *Radio and the Gendered Soundscape*, on the relationship of radio's '"intimate publics" and "imagined communities" 'to 'the development of twentieth-century politics' (p.2).
38. (*The Shadow Spreads) The Shadow of the Swastika 6*, January 11, 1940, 21, Play Library Scripts BBC WAC.
39. (*The Road to War) The Shadow of the Swastika 7*, January 25, 1940, 9, 16, Play Library Scripts BBC WAC.

40. *(The Reichstag Fire), The Shadow of the Swastika 4,* December 14, 1939, 8.15 pm–9 pm, 7, Play Library Scripts BBC WAC.
41. 'Shadow of the Swastika', *Picture Post,* March 2, 1940, 24.
42. See Verma, *Theater of the Mind,* 46.
43. *From War to War,* Thursday February 8, 1940, 27–8, Play Library Scripts, BBC WAC; this section is published at the end of Lloyd and Vinogradoff, 'The Road to War', *Shadow of the Swastika,* 193 with additional directions.
44. Wrigley notes that the female voice also had an important role in MacNeice's wartime features where they served to 'carry energetic political purpose' despite being cast in 'seemingly subsidiary or supporting roles' (Wrigley, "Introduction," 14).
45. *(The Road to War) The Shadow of the Swastika 7,* January 25, 23, 1940, Play Library Scripts BBC WAC. This scene is marked as 'new page' in the PasB script and is not published in Lloyd and Vinogradoff's *Shadow of the Swastika.*
46. *The Listener,* 566, November 16, 1939, 986.
47. *The Listener,* 602, July 25, 1940, 125.
48. R.J.E. Silvey memo to DFD (Laurence Gilliam), 'Listener Research: The Shadow of the Swas-'tika', January 24, 1940. BBC WAC 19/1142/2.
49. *The Shadow of the Swastika* was rebroadcast on the Overseas and Empire Services soon after the Home Service broadcasts, and by the Canadian Broadcasting Corporation (in early 1940); see Nicholas, 151.
50. Laurence Gilliam, "The Job of the Radio Feature Man Today," 4.
51. R.J.E. Silvey memo to DFD (Laurence Gilliam), 'Listener Research: The Shadow of the Swastika', January 24, 1940. BBC WAC 19/1142/2.
52. BBC files show that the early version of the proposed announcement to precede *The Shadow of the Swastika,* which named both Goring as Hitler and Genn as narrator, was replaced with a final version that named only the writers and composer (see drafts in BBC WAC R19/1142/1).
53. Gilliam, "Shadow of the Swastika," *Picture Post,* 24.
54. Ibid.
55. Gilliam, "Shadow of the Swastika," *Picture Post,* 23.
56. Ibid, 25.
57. Lloyd and Vinogradoff, *Shadow of the Swastika,* 11.
58. *(Hitler Over Germany) The Shadow of the Swastika 5,* 12, Play Library Scripts BBC WAC; the musical direction is not included in the printed script (see Lloyd and Vinogradoff, *Shadow of the Swastika,* 118–19).
59. G. Forbes to R. Murray-Leslie, Director of Productions, Pyramid Amalgamated Pictures Ltd., 'Letter from the Deputy Director of the Film Section of the MOI,' February 20, 1940. BBC WAC, R44/173. This file contains further correspondence about proposed film versions of *The Shadow of the Swastika.*
60. *Radio Times,* 867, May 10, 1940, 16.
61. Ibid.
62. Horsley and Tevnan, *Norway Invaded: The First Full Story.*
63. Horsley, *The Battle of Narvik,* 8, Play Library Scripts BBC WAC.
64. Horsley, *The Battle of Narvik,* 10, Play Library Scripts BBC WAC.
65. Goldie, "Critic on the Hearth," May 16, 1940, 987.

66. Frieda Rollo of Essex, letter in *The Times* May 16, 1940, 7.
67. *The Times* May 16, 1940, 7.
68. *The Times* May 17, 1940, 7.
69. 'Extracts from Parliamentary Debate' BBC WAC, R32/44 *P.Q.s Narvik, 1940*.
70. *The Times*, June 12, 1940, 3.
71. *The Times*, June 12, 1940, 3.
72. Gilliam, "The Job of the Radio Feature Man Today," 4.
73. Laurence Gilliam memo to DFD (Geilgud) 'main points to be discussed at Feature Meeting 19th Feb 1941'. BBC WAC R19/352/4.
74. Saunders, *The Battle of Britain*.
75. Campion, *The Battle of Britain*, 94.
76. See Campion, *The Battle of Britain*.
77. D. K. Watson (News Editor *The Daily Mirror*) letter to BBC Press Officer May 9 1941. BBC WAC R19/79/1. This production was rebroadcast on the Forces service on Wednesday 28 May 1941 (9.20–10.20 pm). A recording of the first production can be accessed online http://www.bbc.co.uk/archive/battleofbritain/11425.shtml
78. Goldie, "Critic on the Hearth," August 22, 1940.
79. Nicholas, *The Echo of War*, 201–2.
80. *The Harbour Called Mulberry*, for example, an account of the Normandy landings in June 1944, broadcast on the Home Service Monday March 5, 1945, 9.30–11 pm, was written by McGivern in co-operation with the Admiralty, the War Office, the Ministry of Supply, the Ministry of Labour, the Ministry of War Transport, and the Ministry of Information Films Division and featured only generic characters ('Trooper', 'Contractor', 'Workman', 'Sailor'); see *The Harbour Called Mulberry*, Play Library Scripts BBC WAC and the documents on the production in BBC WAC R19/477.

Bibliography

Bridson, Douglas Geoffrey. *Prospero and Ariel: The Rise and Fall of Radio*. London: Victor Gollancz, 1971.

Briggs, Asa. *The History of Broadcasting in the United Kingdom, Vol. 3. The War of Words, 1939–1945*. Oxford: Oxford University Press, 1995.

Brooks, Tim. *British Propaganda to France, 1940–1944: Machinery, Method and Message*. Edinburgh: Edinburgh University Press, 2007.

Campion, Garry. *The Battle of Britain, 1945–1965: The Air Ministry and the Few*. Houndmills: Palgrave Macmillan, 2015.

Douglas, Susan. *Listening In: Radio and the American Imagination*. Minneapolis: University of Minnesota Press, 2004.

Drakakis, John. "Introduction." In *British Radio Drama*, 1–36. Cambridge: Cambridge University Press, 1981.

Ehrick, Christina. *Radio and the Gendered Soundscape: Women and Broadcasting in Argentina and Uruguay 1930–1950*. Cambridge: Cambridge University Press, 2015.

Gielgud, Val. *Years of the Locust*. London and Brussels: Nicholson & Watson, 1947.

Gilliam, Laurence. "Introduction." In *BBC Features*, 9–14. London: Evans Brothers Limited, 1950.

Gilliam, Laurence. "The Job of the Radio Feature Man Today." *London Calling*, 48. August 8, 1940, 3–4.

Gilliam, Laurence. "Shadow of the Swastika." *Picture Post*, March 2, 1940, 22–25, 53.

Gitelman, Lisa. *Always Already New: Media, History, and the Data of Culture*. Cambridge, MA: The MIT Press, 2006.

Goldie, Grace Wyndham. "Critic on the Hearth." *The Listener*, 592, May 16, 1940, 987.

Goldie, Grace Wyndham. "Critic on the Hearth." *The Listener*, August 22, 1940, 284.

Havers, Richard. *Here is the News: The BBC and the Second World War*. Stroud: Sutton Publishing, 2007.

Horsley, Terence. *The Battle of Narvik*. May 13, 1940. Play Library Scripts, BBC WAC.

Horsley, Terence, and James Tevnan. *Norway Invaded: The First Full Story* (*Cherry Tree War Special*). London & Manchester: Withy Grove Press, 1940.

Lloyd, A. L., and Igor Vinogradoff. *Shadow of the Swastika*. London: Bodley Head, 1940.

Loviglio, Jason. *Radio's Intimate Public: Network Broadcasting and Mass-Mediated Democracy*. Minneapolis: University of Minnesota Press, 2005.

McGivern, Cecil. *Bomb Doors Open and Other Radio War Features*. London: George Allen & Unwin Ltd, 1941.

McHugh, Siobhàn. "The Affective Power of Sound: Oral History on Radio." In *The Oral History Reader*, edited by Robert Perks and Alistair Thomson, 490–507. Abingdon: Routledge, 2016.

Nicholas, Siân. *The Echo of War: Home Front Propaganda and the Wartime BBC, 1939–1940*. Manchester: Manchester University Press, 1996.

Saunders, Hilary Aiden St. George. *The Battle of Britain: An Air Ministry Account of the Great Days from 8th August–31st October 1940*. London: HMSO, 1941.

Sieveking, Lance. *The Stuff of Radio*. London: Cassell, 1934.

Thomas, Jeanette Ann. *A History of the BBC Features Department 1924*–1964. Unpublished thesis submitted to the University of Oxford, 1993.

Verma, Neil. *Theater of the Mind: Imagination, Aesthetics, and American Radio Drama*. Chicago: University of Chicago Press, 2012.

Whitehead, Kate. *The Third Programme: A Literary History*. Oxford: Clarendon Press, 1989.

Wrigley, Amanda. "A Wartime Radio *Odyssey*: Edward Sackville-West and Benjamin Britten's *The Rescue* (1943)." *Radio Journal: International Studies in Broadcast and Audio Media* 8, no. 2 (2011): 81–103.

Wrigley, Amanda. "Introduction: Louis MacNeice, Classical Antiquity, and BBC Radio: From Wartime Propaganda to Radio Plays." In *Louis MacNeice: The Classical Radio Plays*, edited by Amanda Wrigley and Stephen Harrison, 1–30. Oxford: Oxford University Press, 2013.

MAKING WAVES: UNA MARSON'S POETIC VOICE AT THE BBC

Leonie Thomas

Jamaican poet and playwright Una Marson (1905–1965) gained access to the BBC microphone during World War Two and went on to produce, edit, and present an extensive number of broadcasts for the West Indies on the Overseas Service. Closely examining internal BBC communications alongside Marson's poetry, this article asks what it meant for a female colonial subject to be speaking on the radio at this time. Postcolonial theory informs analysis of Marson's poetry before and after her immersion in the metropole. This article focuses attention on the effect of Marson's experiences at the BBC on her literary output and psychological health for the first time. It argues that Marson's interaction with the Corporation profoundly influenced her poetic voice. Furthermore, it uses her poetry to inform readings of archival material that shape our understanding of Marson's relationship to British broadcasting.

The Jamaican poet and playwright Una Marson (1905–1965) became the first black woman employed by the BBC in March 1941. She was appointed as Programmes Assistant on *Calling the West Indies*, the umbrella term for several regular broadcasts to the Caribbean, having freelanced on the BBC's Overseas Service from 1939. This role gave Marson unprecedented access to the BBC microphone as a black, colonial, female subject. Until her departure from the BBC in 1946, Marson wrote scripts, sourced guests, booked bands, edited broadcast messages, and produced high quality radio output for the diverse islands and colonies that came under her remit. Towards the end of her tenure, Marson established the literary magazine programme, *Caribbean Voices*, which would go on to change the trajectory of Caribbean literature in the post-war era. By presenting early editions of this programme, Marson's voice was central to the development of Caribbean literary culture at the BBC. Yet, by the end of 1946, her voice was silenced by a combination of managerial decisions at the Corporation and the onset of severe mental health issues. This article focuses attention on the effect of Marson's experiences at the BBC on her literary output and psychological health for the first time. It argues that Marson's interaction with the Corporation profoundly influenced her poetic voice. Furthermore, it uses her poetry to inform readings of archival material that shape our understanding of Marson's relationship to British broadcasting.

Attention has been paid to the difficulties Marson faced when working at the BBC. James Proctor documents the internal debates that preceded Marson's initial appointment to the Overseas Service as well as the activities she undertook there.[1] As part of this narrative, Proctor explores 'the history of what the director of Empire Services described as "jealousies," "dissentions," and "rivalry" that Marson's appointment provoked'.[2] Marson's experience at the BBC was marked by the competing and sometimes hostile responses of some colleagues and external contacts. This article takes Proctor's work further by analysing the effect of Marson's BBC career on the style, tone, and form of her poetry.

Previous scholarly analysis of Marson's poetry has focussed on her early alienation from Jamaican and British society due to issues of race and gender. Alison Donnell notes that the 'the confounding blend of positions and voices' present in Marson's oeuvre is 'intimately bound to the political conditions of her modernity and alienation as a black woman in both Jamaica and London'.[3] Donnell claims that Marson's critical incongruity in established 'rubrics of modernism' stems from the 'profoundly uneven points of crossing' that structure narratives of modernism, colonialism, and post-colonial literary experimentalism.[4] Donnell's focus, however, is Marson's early poetry and journalism and does not account for the creative influence of BBC broadcasting on Marson's work.

As a Jamaican woman in Britain during the peak of British imperial power, Marson's literary and broadcast output was shaped by her experiences of racism and sexism in the metropole. Anna Snaith relates the impact of these factors on Marson's work, suggesting that Marson employed the 'conventional features and concerns of modernist writing: interiority, urban alienation, technological change and fractured identities and voices'.[5] In other words, Snaith identifies the influence of Marson's experiences in London as a catalyst for experimentalism in the poet's work. Furthermore, the wider contexts of black feminism and black internationalism in which Marson was participating have been noted by Imaobong Umoren, deepening our knowledge of the foundations of Marson's early intellectualism.[6] This article asserts that the internationalist and modernist aspects of Marson's colonial experience were exacerbated by her BBC work.

Indeed, broadcasting's transnational aspects can inform our understanding of Marson's poetry. Speaking in London but heard by audiences in the West Indies, Marson challenges established narratives of the way modernism circulated from the centre to the periphery. In Laura Winkiel's words, a transnational critical focus problematises 'a hierarchical form of modernism said to emanate from the European metropolis and travel to the colonies'.[7] Marson herself, and many of the contributors on *Calling the West Indies,* had travelled from the colonies to London, and would continue to move between centre and margin according to the demands of war and the opportunities presented by peace. Such a destabilisation of the hierarchies of modernism enables scholars to trace what Mary Lou Emery has described as the 'complex relay of cultural influences' that offers understanding of 'the cultural formation of colonial subjectivities'.[8] Thus, the broad range of cultural influences that inspired Marson can be laid bare through a transnational lens.

Building on these critical conversations, this article considers the pastoral aspects of Marson's early poetry, composed before she visited London in 1932. These works are analysed through a postcolonial framework of mimicry, the appropriation of imperial language by colonised peoples, as theorised by Homi Bhabha. This framework allows us to chart the development of Marson's poetic voice as she engaged with strains of imperialist discourse in the metropole. A consideration of the role and influence of creole in Marson's later poetry will follow to explore the vocal experimentation that colonial consciousness had on Marson's poetics. Finally, archival evidence from the BBC Written Archive Centre reveals that Marson's time at the BBC caused her to develop a more experimental poetic persona. This encounter, however, was complex. Whilst the BBC provided a creative and galvanising force for her poetry, the writer's time at the Corporation also saw a deterioration of her mental health. In January 1946, Marson left the BBC due to the onset of depression and what was diagnosed as schizophrenia, never to publish or broadcast again. Thus, her poetry provides an opportunity to track the trajectory of Marson's voice across her broadcasting experience and suggests the problematic influence of the BBC as a creative contact zone and as a site of exposure and vulnerability for this black, female, colonial writer.[9]

Poetic Voice in a Colonial Context

Marson published her first book of poems, *Tropic Reveries*, in Jamaica in 1930, followed by *Heights and Depths* in 1931. The poems of these collections are vivid, youthful, sometimes unadventurous, and sometimes forthright. Donnell writes of her early work, 'the fact that it is formally tame does not prohibit the stirrings of a politically adventurous voice'.[10] Donnell hears the strains of potentiality in Marson's earlier work, connecting her poetic voice to the politically brave voice that, later in the decade, would be heard in a secretarial role at the League of Nations, at the International Alliance of Women for Suffrage and Equal Citizenship Conference in 1935, and on the airwaves of the BBC from 1939. Furthermore, for the purposes of this argument and its movement across various literal and theoretical instances of Marson's spoken, broadcast, and poetic utterance, voice is defined as the 'qualities of vocal sounds [that] enter directly into the aesthetic experience'.[11] This definition foregrounds the ability of vocal signifiers to transcend specific forms or modes.

Marson's early poetry, written before her journey to London in 1932, expresses a colonial perspective of Jamaica which conforms to 'traditional' poetic conventions. In 'Jamaica', published in *Tropic Reveries*, Marson evokes a pastiche of an English summer:

> The hum of bees upon the flowering trees,
> Makes sweetest music with the summer breeze.
> The fields are covered o'er with Daisies bright
> Which nod their pretty heads in sheer delight[12]

The clichés are tired; bees and daisies humming and nodding with boredom. Although Emery suggests that Marson used lyric and sonnet forms to 'implicitly subvert sexist and colonialist hierarchies', the uncritical use of pastoral imagery here reasserts colonialist hierarchies by describing an island colony in over-determined, feminised language.[13] In the final stanza, the island itself calls out to its potential visitors, 'Come, be my guest/And I will give thee gladness, peace and rest!'.[14] This celebratory exclamation appears to be addressed to the white plantation classes, the only inhabitants of the island who are free from toil. The sentence is written in an archaic English that conforms to expectations of poetic voice, the old fashioned 'thee' bolstered by the triad of wholesome virtues 'gladness, peace and rest'. Similarly, a second poem called 'Jamaica', published in *Heights and Depths*, offers a reductive and imperially tinged rendering of the island: 'Just a lovely little jewel floating on fair Carib's breast'/All a-glittering in her verdue 'neath a blazing tropic sky'.[15] The characterisation of 'jewel' evokes discourses of imperial jewels in the crown. The passive verbs 'floating' and 'a-glittering' corroborate a colonising worldview where land is ripe for the taking and native opposition is effectively silenced.

Why, then, does Marson's voice chime so strikingly with an imperial perspective in these early poems about Jamaica? In part, perhaps, because she had hardly heard anything else. Donnell notes:

> These poems are ripe with internal contradictions and make visible the ways in which the literary quest to represent an 'authentic' Jamaica was already compromised and entangled by the prestige models of colonial literary landscapes that created constant recourse to culture, even for representations of Nature.[16]

The 'prestige models of colonial literary landscapes' were superimposed onto the Jamaican topography. In addition to racial hierarchies structuring economic advantage across the region, the West Indies was subjected to British cultural values. Jamaican schools taught English poetry and English was the dominant language of the region. The diverse cultures of the indigenous people of the Caribbean, as well as the customs of African slaves and indentured workers from the Indian subcontinent, were diminished and disregarded as a result of British colonial educational policy. Caribbean poet Philip Sherlock writes in the foreword of the first anthology of West Indian poetry, published in 1966: 'Without knowing why, I felt a little poverty-stricken, a poor relation, because there was no West Indian poet with whom I might walk as I did the poets of England'.[17] West Indian poets were taught the cadences of British poetry and were denied recognisable West Indian voices.

Thus, Marson's early poetic voice is infused with stylistic imitation, an attempt to find a poetic persona when the cultural norms were dominated by the values and voices of the coloniser. Postcolonial theorist Homi Bhabha articulates the ideological implications of such mimicry in a colonial context: 'Colonial mimicry is the desire for a reformed, recognisable Other, as *a subject of a difference that is almost the same, but not quite*'.[18] The power of mimicry in colonial discourse arises from the ambivalence that is produced when the almost-the-same-but-not-quite-ness of the interpolated colonial Other becomes apparent. Bhabha describes the ways in which the '*ambivalence* of mimicry … does not merely "rupture" the discourse, but becomes transformed into an uncertainty which fixes the colonial subject as a "partial" presence'.[19] In other words, the uncertainty that is produced by a colonial writer using imperial forms of literature creates an ambivalence in their work that prevents it from becoming literarily complete, whole, or successful by Western standards. The colonial subject/writer becomes fixed as a 'partial' presence. Graham Huggan notes the way mimicry enables a colonial writer to '[take] up the metropolitan desire to hear the strains of [their] own voice' before '[rearticulating] that desire' as almost the same but not quite.[20] In this framework, Marson's pastoral English poetry troubles literary convention by describing a fundamentally different landscape—a tropical climate—whilst using typically Western devices.

One way of addressing the ambivalence of mimicry is through the use of Kamau Braithwaite's concept of 'nation language', embracing colony-specific variants of the English language as a method for articulating a colonial voice more fully.[21] Braithwaite describes English in the Caribbean as 'not the standard, imported, educated English, but that of the submerged, surrealist experience and sensibility, which has always been there … an English which is like a howl, or a shout or a machine-gun or the wind or a wave'.[22] The colonial subject therefore, contrapuntally, uses a fractured language to establish a wholeness, a completion that rejects imperial cultural hierarchies. The *Oxford English Dictionary* elides 'nation language' with the linguistic term 'creole', describing it as:

> A language that has developed from the mixing of two or more parent languages … typically arising as the result of contact between the language of a dominant group … and that of a subordinate group.[23]

Although this definition does not reflect the full complexity of the linguistic and cultural debates that surround this term, it makes clear the power dynamics associated with creole.[24]

Indeed, it was not until Marson experienced the power and prejudice of the metropole that she began using West Indian speech patterns in her poetry. In *Black Skin, White Masks* (1952), Frantz Fanon explores the ideological implications of a colonial subject for adopting an imperial language. Fanon argues, 'The black man is unaware of it as long as he lives among his own people; but at the first white gaze, he feels the weight of his melanin'.[25] It was only *after* Marson's initial visit to Britain and her subsequent return to Jamaica in 1936 that she began experimenting with creole in her poetry. For example, creole is used to represent the conversation of a pair of female labourers in 'The Stone Breakers', published in the 1937 collection, *The Moth and the Star*. The women share their grievances:

> De groun' is dat dry,
> Not a ting will grow –
> Massy Lard, dis life is hard.[26]

The hard sounds of the 'd' and 't' consonants rumble around the verse like the stones that the women must break each day, the non-standard formation of 'De', 'dis', and 'ting' evoking the subterranean aspects of creole that Braithwaite highlights. Marson's poetic voice diversifies after her first experience of the imperial metropole.

Moreover, Elaine Savory notes the complexity of moving between multiple vocal registers for West Indian writers:

> the music of West Indian poetry, its rhythms and rhymes and line breaks and stanza forms, is always complex, always working something out between different but related systems, between the traditions and models of the scribal and the always revisioning performative identity of the oral.[27]

West Indian poetry's movement between scribal and oral forms of language, and its quality of 'working something out' is also experienced by the radio broadcaster. Andrew Crisell asserts that 'words on the radio could be regarded as the application of oral language to a situation which normally calls for writing', the situation being the separation of sender and receiver.[28] Furthermore, in the 1940s, radio broadcasts were scripted, meaning that a scribal form was used to shape an oral delivery. Radio words are further enriched and given meaning through the voice and its representation of the person, the network, or even the nation that the voice conjures.[29] It is this added weight of representation that makes Marson's appearances at the BBC's microphone so significant because she is combing the tensions of a West Indian poetic voice with the experience of radio's contradictory oral/scribal demands. In 1939, having experimented with different vocal registers and modes in the creole of her poetry, Marson's voice was about to be exposed to further modernist influence in the studios of the BBC.

On Air at the BBC Overseas Service

In 1932, the BBC had introduced an Empire Service with the remit of catering to white Britons abroad. At the outbreak of war, the strategic importance of broadcasting to non-white colonial subjects to mobilise them for the war effort became apparent. As scholars of the BBC World Service assert, '[t]he strategic need to ensure that listeners should be made aware of the British interpretation of events and not be dependent on the views of Italian and German international broadcasting ... was considerable'.[30] In 1939, to address these concerns, the Empire Service became the Overseas Service and was accompanied by a significant expansion of coverage and output, funded by Foreign Office grant-in-aid money.[31] This, in turn, sparked a recruitment drive at the BBC for people who had knowledge of the extensive geographical regions that were now being addressed, causing an influx of diverse voices and perspectives into the Corporation. Una Marson's voice was one of them.

As a result of the increased diversity of staff at the Service, there were many examples of the BBC functioning as a creative and liberating facilitator of cross-cultural exchange. Most famously, in November 1942, the writers George Orwell, Mulk Raj Anand, T. S. Eliot, Una Marson, Venu Chitale, M. J. Tambimuttu, Narayana Menon, and William Empson gathered for the BBC programme *Voice,* a six-part poetry magazine programme produced by Orwell and broadcast to India on the BBC Eastern Service. Marson's inclusion in this moment cements her place in the currents of transnational modernism that were beginning to be felt. During the programme, Marson read from 'The Banjo Boy' (*The Moth and the Star*, 1937):

Where did you get that rhythm?
That swing and that motion,
That bubbling laughter
With which you punctuate
Your songs? I have it too,
I can feel it going through me,
But I can't express like you do.[32]

Here, the embodied rhythm of 'swing' and 'motion' are halted by the relative clauses 'that', pinned in the context of aesthetic comparison. In this poem, Marson compares her lyrical skills with the Banjo Boy ('I have it too'), in the same way that Marson's poetry was compared with the other poets on the programme. The structure of the *Voice* programme meant that comparative analysis and critical engagement sat alongside featured poems. As such, Marson was participating in the same critical experience as Eliot, her 'The Banjo Boy' broadcast alongside *The Waste Land* and subjected to the same scrutiny. This was an example of vocal equality between canonical modernists, such as Eliot, and colonial poets, such as Marson—they shared the same microphone.

Importantly, Marson's display of stylistic experimentalism on *Voice* foreshadows the well-documented history of the arrival of Caribbean intellectuals in London during the 1950s and their association with elite modernism, as outlined by Peter Kalliney. Kalliney argues that:

> By inviting colonial intellectuals to help revive and reshape cultural institutions during the mid-century period, metropolitan modernists hoped to preserve the aesthetic temper of interwar literature while expanding its geographical reach.[33]

In light of growing concerns about the waning influence of British modernism brought about by a perceived dearth of experimental young writers in the 1930s and 1940s, metropolitan literary intellectuals sought 'new lines of aesthetic exchange' with writers from the crumbling empire.[34] Indeed, part of Marson's modernity lies in the fact that she was participating in broadcast and cultural networks *before* many other Caribbean writers.[35] The final line of 'The Banjo Boy' gestures towards the difficulty of expressing herself in the hostile metropole before more solid networks of aesthetic exchange had been established: 'But I can't express like you do'.[36] Foregrounding the contrast between the poetic persona and the creative power of the banjo boy with the conjunction, 'but', Marson articulates the difficulty that faced colonial subjects in their self-articulation.

The question of genuine cultural inclusion and its relationship with verbal alterity can be seen in Marson's later poetry. In 'The Stranger' (1937), Marson illustrates how a colonial voice is subjected to British scrutiny. A West Indian woman talks to a British man in London and is courted for her exotic stories of the colonies:

> You liked talking to people like me
> You said, with a wistful smile
> That enchanted me, so the pause
> That came before I spoke
> Must have seemed strange to you[37]

Here, Marson challenges the speech pattern of Standard English by making her run-on lines alter the emphasis for the reader. The caesura before 'so the pause' works to keep the reader in suspense and the run-on line after 'people like me' leaves the racially loaded accusation of stereotyping hanging in the air. The power of the reported speech, 'you said', gives vocal agency to the subaltern persona.

It was reported speech and the repetition of criticism made by others that impinged on Marson's career at the BBC. On 6 March 1942, J. B. Clark, African Controller of the Overseas Service, received a letter from Harold Nicolson, then Member of Parliament and BBC governor. The letter outlined criticisms made by the West India Committee, a prominent and politically active organisation that had received a Royal Charter for its welfare work in the Caribbean but still represented predominantly white, planter interests. Their London base, Aggrey House, acted a social hub for West Indians arriving in Britain to support the war effort. The Committee was hostile to Marson for suggesting non-white West Indians be employed at Aggrey House and had frustrated Marson's attempts to represent the diversity of the West Indian population in her programmes, arguing for an unrepresentative 50:50 ratio of white and non-white speakers.[38] Although archival files in the BBC do not hold details of the criticisms Nicolson reported, Clark replied that 'there is a good deal of evidence of both personal and colour prejudice,' from the Committee.[39] Clark asserted that racism played a significant part in sparking the criticism and refused to take the matter further.

Although Clark's response conveyed coherent BBC policy to Nicolson, the question of the Corporation's handling of racism was by no means a settled matter within the Overseas Service. In a report to Clark on the issue, Marson's manager and the Director of the African Service, John Grenfell Williams was explicit about the issues faced:

> [A]s long as we have Miss Marson or any other coloured West Indian on the staff we shall be confronted with this kind of problem from time to time ... Miss Marson is the only coloured West Indian in this country capable of handling these programmes. And ... with the limited material at her disposal and the extremely hard work involved she has done a very good job.[40]

Grenfell Williams was adamant that Marson was the right woman for the job and, when prejudice was an expected consequence of employing a black West Indian, all unfounded criticism should be ignored. Grenfell Williams is more generous than Clark in his defence of Marson, noting the 'limited material' and 'extremely hard work' that she had put in to make the programme a success.

Moreover, the prominence of Marson's voice, and its specific Jamaican intonation, drew complaints from other external sources. In January 1943, Sir Frank Stockdale, the Comptroller for development and welfare of the West Indies, had contacted the BBC with concerns about Marson. Stockdale was responsible for distributing Colonial Office funds to the region, with a focus on cultural development and agriculture. Thus, his criticism that *Calling the West Indies* featured '"Too much Jamaica, too much Una Marson"' was of serious concern to BBC staff because his support was vital for improving broadcasting infrastructure and increasing listenership in the region.[41] The accusation was broadly rebuffed by the BBC as a necessary by-product of the limited resources and personnel available to them but it was repeated throughout Marson's career.

In August 1944, for example, Grenfell Williams raised more of Stockdale's criticisms with Marson and recorded it in a report: 'I told Miss Marson that Sir Frank Stockdale had reported that people in the West Indies were critical of her regular and frequent appearances at the microphone, and that some feeling against her was growing'.[42] Marson defended herself in the exchange, stating that:

> She had never come across this particular criticism among the many hundreds of West Indian serving men and women and factory workers with who she has worked in England ... a large number of the coloured ones, representative of all islands, had gone out of their way to tell her how proud they were that a coloured West Indian was appearing in the BBC programmes.[43]

Marson tries to further the opinion of the large, non-white audience, with whom she had extensive contact as a contrast to the perspective of a colonial official. Marson's voice itself is reported by BBC staff and therefore effaced by administrative systems of management. By the end of the report, Marson's exhaustion is clear: 'She added that she was extremely tired and would like to ask again whether the Corporation would consider allowing her to go to the West Indies for a few months at the end of the year'.[44] Marson's voice, as non-white and colonial, is marginalised by the process of BBC administration in much the same way that the non-white West Indians' opinions are outweighed by one colonial official, demonstrating the colonial hierarchy that structured questions of voice on the airwaves.

Later in August 1944, R. A. Rendall, Assistant Controller of the Overseas Service, outlined the consequences of Stockdale's criticisms in a memo to Grenfell Williams:

> I think the truth of the matter is that no one person can successfully sustain the position in which Miss Marson has been placed, and I think that she should be made to concentrate on organisation and that her appearance at the microphone be irregular and rare; nor should her name feature in credits as producer, organiser, editor etc.[45]

The 'position' in which Marson had been placed was indeed one of significant pressure from audiences and colonial officials, coupled with an increasing lack of support from BBC staff. When Rendall advised Marson to keep her 'appearances at the microphone ... irregular and rare', as well as removing her name from the credits, he effectively marginalised her voice, revoking her access to the microphone and denying her appropriate recognition for her work. Rather than choose to ignore the racism that seemed inevitable to Grenfell Williams in 1942, Rendall acts to silence Marson.

Mimicry and Modernism

Marson's most modernist poetry coincides with this silencing of her broadcast voice and the decline of her mental health. Due, in part, to the severity of her illness, no documents or evidence survive of Marson's experience of the end of her tenure at the BBC. As such, it is only her poetry that can give us an insight into her perspective during this time. Her final book of poetry, which reprints many poems that appeared in her previous three volumes, was published by the University of London Press in 1945 under the title *Towards the Stars*. Contained within that collection are new poems of starkness and desolation. They tell the story of a woman struggling to speak out for what she believed in, to shout into the abyss of homesickness and racism that surrounded her. In the poetry she wrote after 1942, the creole-based verbal markers of alterity that Marson initially embraced as part of her developing modernity are replaced by staccato syllables that crackle on the page. In particular, the poem 'Politeness' bristles with inexpressible rage, drawing the lines of tension and the lacunae of misunderstanding and racism in broad strokes:

> They tell us
> That our skin is black
> But our hearts are white.
> We tell them
> That their skin is white
> But their hearts are black.[46]

Marson articulates the insurmountable differences between 'them and us' through the politically loaded verb 'tell'. 'Tell' works to evoke the paternalistic tones of the white colonisers, 'reassuring' those black visitors to Britain that they are almost, but not quite, white. However, Marson's poem turns on the plural pronoun 'we', claiming the ability to speak, to 'tell', colonisers that their racism and oppression is cruel. Telling is not mimicry; telling is articulation, speaking back to empire with full force and self-fulfilled power.[47] This six-line poem's impact lies in its familiarity, frequency, and crushing inescapability, refracted through the distorting stylistic experimentalism of Marson's modernism.

This pinnacle of Marson's modernist voice aligned with her creation of a broadcast platform for Caribbean writers. There is some critical uncertainty about the exact starting date of *Caribbean Voices.* Proctor suggests the earliest date of possible broadcast of the influential programme was 1 October 1944, and script indexes show that the title 'Caribbean Voices' was being used as early as 3 December 1944, although whether this indicates a segment of the broader *Calling the West Indies* show or a whole programme is unclear.[48] What Proctor highlights is 'just how closely the inauguration *Caribbean Voices* coincided with the request for Marson to step back from the microphone (August 1944), her bouts of leave, and her subsequent breakdown the following year'.[49] Marson's exclusion from the microphone under Rendall's direction coincides with the point at which Caribbean writers gained greater access to the machinery of the BBC.

Yet it was Marson's voice itself that framed those early editions of *Caribbean Voices.* As compere and producer of the programme, Marson offered 'short linking material for the talk', establishing connections between the West Indian literature that was featured.[50] On Sunday 25 March 1945, for example, Marson presented a *Caribbean Voices* tribute to the recently deceased Jamaican poet, Constance Hollar. Marson is reverential: 'No words of mine can adequately describe my personal loss, and the loss that we in the West Indies have sustained in the passing of Constance Hollar'.[51] Marson unites the audience, 'we in the West Indies', in appreciating and mourning the poet, playing with the contradiction of denying her voice, 'no words of mine', whilst actively speaking on the radio. Furthermore, one of Marson's own poems, 'Towards the Stars' (1945) was broadcast in this programme but attributed, vaguely, to 'a Jamaican away from home'.[52] The contradictory positioning of Marson as articulate poet and abstract exile complicates her relationship with her own Caribbean voice. In a programme dedicated to a female poet, with all readings provided by West Indian women and introduced by Marson herself, Marson does not explicitly claim her own voice due to the administrative restrictions that had been placed on her.

Finally, it is alienation from her own voice that characterises the account of Marson's declining mental health in the BBC archives. The managerial and bureaucratic reporting of Marson's perspective that had recurred during her employment (the reports of conversations by Rendall and Grenfell Williams, for example) now escalated. Despite journeying to Jamaica in July 1945 for rest leave, Marson was not fit and well on her return to the office in December 1945. Her colleague and friend, Joan Doulton, records Marson's deterioration in a memorandum. On 28 December 1945, '[Marson] was seen by a specialist Dr Noel Harris, who reported serious nervous breakdown and advised her to go into a Nursing Home'.[53] The report goes on to detail Marson's refusal of treatment and eventual certification. Due to the BBC's duty of care to Marson, Doulton herself signed the document that committed Marson to hospital. Furthermore, BBC medical advisor, Dr A. F. Whyte reported in May 1946 that, 'Miss Marson is suffering from schizophrenia and has delusions of persecution by a group of individuals, but she is unable to say who the members of this group are'.[54] Whyte's phrasing of his diagnosis and specific description of Marson as 'unable to say' who was tormenting her demonstrates the extent to which Marson was silenced at this point. Although the BBC continued to pay her full salary for ten months during her hospitalisation, Marson did not recover in England and was sent back to Jamaica in the company of her friends in October 1946: she never published another poem and she never broadcast again.

The trajectory of Marson's voice, both poetic and broadcast, over the course of her writing career can be seen to move from mimetic and pastoral, through to creolised and racially conscious, and finally to prominently modernist and outspoken. At this point of creative articulation, Marson's mental health issues led to her effective silencing, her loss of voice at the BBC and forfeiture of agency. Managerial decisions at the BBC had resulted in a decrease of Marson's access to the microphone, although she was still audibly involved in the development of *Caribbean Voices*. On the airwaves of the BBC, Marson made strides to provide a platform for Caribbean writers to explore their literary voices. In the process of this, she lost her own.[55]

Disclosure Statement

No potential conflict of interest was reported by the author.

Notes

1. Proctor, "Una Marson at the BBC".
2. Proctor, "Una Marson at the BBC," 4.
3. Donnell, "Una Marson and the Fractured Subjects of Modernity," 346.
4. Donnell, "Una Marson and the Fractured Subjects of Modernity," 346–48.
5. Snaith, *Modernist Voyages*, 154.
6. Umoren, "'This is the Age of Woman'," 58–67.
7. Winkiel, "Gendered Transnationalism," 40.
8. Emery, *Modernism*, 119.
9. I use the term 'colonial' here to acknowledge the Crown Colony status of Jamaica during this period, making Marson's citizenship administratively colonial, but also to acknowledge her position as a late-colonial intellectual as subject to imperial ideology and oppression *before* the nationalist movements of the mid-century and subsequent independence.
10. Donnell, "Introduction," 13.
11. Preminger and Brogan, *Poetry and Poetics*, 1366.
12. Marson, "Jamaica", *Selected Poems*, 56.
13. Emery, 113.
14. Marson, "Jamaica," 57.
15. Marson, "Jamaica," 73.
16. Donnell, "Introduction," 29.
17. Sherlock, "Foreword," vii.
18. Bhabha, "Of Mimicry and Man," 126.
19. Bhabha, 127.
20. Huggan, "A Tale of Two Parrots," 645.
21. Braithwaite, *History of the Voice*, 5.
22. Braithwaite, 13.

23. Oxford English Dictionary Online, *s.v.* 'Creole, n. and adj.', http://www.oed.com/view/Entry/44229?redirectedFrom=creole.
24. See McWhorter, *Defining Creole*, for an overview of the controversies in the field of linguistics.
25. Fanon, *Black Skin, White Masks*, 128.
26. Marson, "The Stone Breakers," 125.
27. Savory, "Anglophone Caribbean Literature," 729.
28. Crisell, *Understanding Radio*, 58.
29. Crisell, 47.
30. Gillespie and Webb, "Corporate Cosmopolitanism," 5.
31. For further details of the changed relations between the BBC and the wartime government, and the consequences of such wartime collaboration on Empire broadcasting, see Potter, *Broadcasting Empire*, 114.
32. Marson, "The Banjo Boy," 124.
33. Kalliney, *Commonwealth of Letters*, 4.
34. Ibid.
35. Rosenburg discusses Marson's position between interwar Caribbean writers and the prominent writers of the 1950s. Rosenburg, *Nationalism and the Formation of Caribbean Literature*, 159–80.
36. Marson, "The Banjo Boy," 124.
37. Marson, "The Stranger," 98.
38. Proctor, 14–15.
39. J. B. Clark, letter to Harold Nicolson, 7 March 1942, E2/584, Foreign Gen. West Indies, 1939–1950, BBC Written Archive Centre (hereafter BBC WAC).
40. John Grenfell Williams, letter to J. B. Clark, 11 March 1942, E2/584, Foreign Gen. West Indies, 1939–1950, BBC WAC.
41. Rendall, letter to Grenfell Williams, 2 January 1943, E1/1294/2 (File 1b, 1943–1944), Countries: West Indies, Broadcasting in West Indies. BBC WAC.
42. Grenfell Williams to Rendall, "Notes on Interview with Miss Marson," 28 August 1944, L1/290/1, BBC WAC.
43. Ibid.
44. Ibid.
45. Rendall, letter to Grenfell Williams, 7 August 1944, L1/290/1, BBC WAC.
46. Marson, "Politeness," 166.
47. A modified instance of the influential concept of 'writing back to Empire' as theorised by Ashcroft, Griffiths and Tiffin in *The Empire Writes Back*, 1989.
48. Proctor, "Una Marson," 24. and *Calling the West Indies* script in BBC WAC.
49. Proctor, "Una Marson," 24.
50. Grenfell Williams to Rendall, "Notes on Interview with Miss Marson," 28 August 1944, L1/290/1, BBC WAC.
51. *Caribbean Voices*, 25 March 1945, scripts in BBC WAC.
52. Ibid.
53. Undated memorandum written by Joan Doulton, detailing the events of Marson's illness, communication with her sister, and dates of movement between hospitals, L1/290/1, BBC WAC.

54. Report from A. F. Whyte, BBC Medical Advisor, titled 'Miss Una Marson', 3 May 1946, L1/290/2, BBC WAC.

55. See Newton, *Paving the Empire Road*, for the inclusion of Black Britons at the BBC.

Bibliography

Ashcroft, Bill, Gareth Griffiths, and Helen Tiffin. *The Empire Writes Back: Theory and Practice in Post-colonial Literatures*. London and New York: Routledge, 1989.

Bhabha, Homi. "Of Mimicry and Man: The Ambivalence of Colonial Discourse." *Discipleship: A Special Issue on Psychoanalysis* 28, no. 1 (1984): 125–133.

Braithwaite, Kamau. *History of the Voice: The Development of Nation Language in Anglophone Caribbean Poetry*. London: New Beacon Books, 1984.

Crisell, Andrew. *Understanding Radio*. London and New York: Methuen, 1986.

Donnell, Alison. "Una Marson and the Fractured Subjects of Modernity: Writing across the Black Atlantic." *Women: A Cultural Review* 22, no. 4 (2011): 345–369.

Donnell, Alison. "Introduction." In *Una Marson: Selected Poems*, edited by Alison Donnell, 11–39. Leeds: Peepal Tree Press, 2011.

Emery, Mary Lou. *Modernism, the Visual, and Caribbean Literature*. Cambridge: Cambridge University Press, 2007.

Fanon, Frantz. *Black Skin, White Masks*. Translated and edited by Richard Philcox. New York: Grove Press, 2008.

Gillespie, Marie, and Alban Webb. "Corporate Cosmopolitanism: Diasporas and Diplomacy at the BBC, 1932–2012." In *Diasporas and Diplomacy: Cosmopolitan Contact Zones at the BBC World Service (1932–2012)*, edited by Marie Gillespie and Alban Webb, 1–20. London: Routledge, 2013.

Huggan, Graham. "A Tale of Two Parrots: Walcott, Rhys, and the Uses of Colonial Mimicry." *Contemporary Literature* 35, no. 4 (1994): 643–660.

Jarrett-Macaulay, Delia. *The Life of Una Marson, 1905–1965*. Manchester: Manchester University Press, 1998.

Kalliney, Peter J. *Commonwealth of Letters: British Literary Culture and the Emergence of Postcolonial Aesthetics*. Oxford: Oxford University Press, 2013.

Marson, Una. *Una Marson, Selected Poems*. Edited by Alison Donnell. Leeds: Peepal Tree Press, 2011.

McWhorter, John H. *Defining Creole*. Oxford: Oxford University Press, 2005.

Morris, Mervyn. *Is English We Speaking and Other Essays*. Kingston: Ian Randle, 1999.

Newton, Darrell M. *Paving the Empire Road: BBC Television and Black Britons*. Manchester: Manchester University Press, 2011.

Potter, Simon. *Broadcasting Empire: The BBC and the British World, 1922–1970*. Oxford: Oxford University Press, 2012.

Preminger, Alex, and T. V. F. Brogen. *The New Princeton Encyclopedia of Poetry and Poetics*. Associate Editors Frank J. Warnke, O. B. Hardison and Earl Miner. Princeton: Princeton University Press, 1993.

Proctor, James. "Una Marson at the BBC." *Small Axe* 19 (2015): 1–28.

Rosenberg, Leah Reade. *Nationalism and the Formation of Caribbean Literature*. New York: Palgrave Macmillan, 2007.

Savory, Elaine. "Anglophone Caribbean Literature." In *The Cambridge History of African and Caribbean Literature*, edited by F. Abiola Irele and Simon Gikandi, Vol. 2, 711–758. Cambridge: Cambridge University Press, 2004.

Sherlock, Philip. "Foreword." In *Caribbean Voices: An Anthology of West Indian Poetry*, edited by John Figueroa, Vol. 1, vii–viii. London: Evans Brothers, 1966.

Snaith, Anna. *Modernist Voyages: Colonial Women Writers in London, 1890–1945*. Cambridge: Cambridge University Press, 2014.

Umoren, Imaobong D. "'This is the Age of Woman': Black Feminism and Black Internationalism in the Works of Una Marson, 1928–1938." *History of Women in the Americas* 1 (2013): 50–73.

Winkiel, Laura. "Gendered Transnationalism in 'The New Modernist Studies'." *Literature Compass* 10 (2013): 38–44.

'COUNTRIES IN THE AIR': TRAVEL AND GEOMODERNISM IN LOUIS MACNEICE'S BBC FEATURES

Aasiya Lodhi

In the middle stretch of his twenty-two-year BBC career, the poet and producer Louis MacNeice earned a reputation as one of the 'undisputed masters of creative sound broadcasting', a reputation derived, in part, from a huge range of radio features that were founded upon his journeys abroad. Through close examination of some of his most significant overseas soundscapes—including Portrait of Rome *(1947) and* Portrait of Delhi *(1948)—this article will consider the role and function of travel in shaping MacNeice's engagement with the radio feature as a modernist form at a particular transcultural moment when Britain moved through the end of the Second World War and the eventual disintegration of its empire.*

Introduction

Writing to T. S. Eliot in 1957, the writer and radio producer Louis MacNeice pitched an idea for a 'not too usual kind of travel book',[1] a volume comprised of his BBC Features scripts anchored in travels abroad.[2] MacNeice's proposed title, 'Countries in the Air', emphasised the twinning of his radio craft with journeying overseas—a crossing of continents via airwaves and air travel—that imprinted a trail across his broadcast output in the 1940s and 1950s.[3] Growing out of the transatlantic and European focused programmes on which MacNeice cut his teeth as a wartime writer and producer from 1941, these post-war broadcasts mark a departure in his feature-making through an explicit engagement with the concept of going abroad, drawing directly on time spent on the ground and signposted to the listener often through the use of a central character in the guise of tourist. Heading into the 'middle stretch'[4] of his two-decade radio career, from the end of the 1940s to the mid-1950s, MacNeice, one of the leading lights of the BBC's Features Department, melded travel and the radio feature form to bear witness to nascent independent nationhood against the backdrop of crumbling empire and to reinterpret non-British history for domestic audiences, often utilising the cityscape as recurrent trope.

Through a focus on three of his overseas features – with close analysis of *Portrait of Rome* (1947) and *Portrait of Delhi* (1948), and brief reference to *Portrait of Athens* (1951) – this article explores how travel and geography intersect with conceptualisations of spatiality and the radio feature form in MacNeice's productions to combine into a type of radio 'geomodernism'. As Laura Doyle and Laura Winkiel contend, geomodernist texts 'signal [s] a locational approach' to modernity through a heightened awareness of national and geographical boundaries and their mutability, and by an engagement with notions of landscape, terrain and space.[5] Tracing these strands in MacNeice's post-war travel programmes reveals the contours of a geographically oriented modernism constituted through the radio feature. MacNeice's aesthetics display a reflexive awareness about broadcasting's abilities to cross national borders and to reimagine space and place through the broadcast vernacular. They also serve to highlight how the BBC interpreted a changing global landscape, through the grammar of radio, at a particularly significant transcultural moment when Britain moved through the end of the Second World War and the eventual disintegration of its empire.

Radio's Bi-directionality

Radio's overseas broadcasting had been explicitly embedded in the BBC's strategy at least from 1932 onwards with the launch of its Empire Service. The network sat alongside the Corporation's small but significant foreign-themed domestic programming, which included content intended for those considering emigration, such as *Girls and A Career Overseas* (National Programme, 12 November 1930), produced under the auspices of the Overseas Settlement Department. In addition, British listeners at home were offered output that reflected back imperial projections of Britishness abroad though the Christmas and Empire Day relay broadcasts, begun in the 1920s.[6] But they also learned of copper in Belgian Congo, coal mining in Western Pennsylvania, and of life in the Amazon rainforest in programmes such as *Travel Talk,* a series that foregrounded journeying to shed light on world cultures from the viewpoint of a British listening public. The series, however, also pointed to possibilities and potentialities—of cultural exchange, influence and permeation—that could be set in motion by travel.[7] In an edition on Warsaw that explored the 'medieval and romantic' nature of the Polish people (8 September 1927), Val Gielgud, later to become Head of the Drama Department,[8] drew on his Polish background to 'visit the country of his blood' whilst acknowledging its influence on his own creative output.[9] Gielgud's outwards-inwards programme serves as a reminder of what Michele Hilmes has described as the 'inherent *transnationalism* of broadcasting's cultural economy', in which 'other influences stream into the national space' through the networks and channels of national broadcasting.[10] Hilmes is concerned with the relationship between Britain and America but her conceptualisation of national broadcast organisations as 'circulatory systems',[11] reliant on elements of bi-directionality, provides a useful lens for the examination of domestic BBC Radio's engagements with travel, especially at the end of the decade that followed the launch of *Travel Talk* when wartime cemented overseas programming and 'other influences' into the BBC's home schedules.[12]

Features Department in particular—merged with the Drama Department during wartime—brought a strong international sensibility to its output for the Home Service and later the Third Programme, through long-running series such as *Window on Europe* (first aired during wartime and continuing into the 1950s).[13] Features producers Rayner Heppenstall and Geoffrey Bridson recounted in their memoirs the many travels they were encouraged to make to gather programme source material across Europe but also much further afield, and the globe-crossing that was committed to and funded by the BBC in the postwar period.[14] Of the fifteen scripts chosen for print publication in a special anthology in 1950 by Features head Laurence Gilliam, nine were directly associated with overseas subjects and the potted biographies of the writer-producers responsible revealed an enormously well-travelled and globally-minded team.[15] Leonard Cottrell, for example, had 'flown many thousands of miles in search of feature material', including to Kenya, Tanganyika, and the West Indies; Alan Burgess, wrote Gilliam, had wandered and 'hoboed' in his pre-BBC life through the South Sea Islands, Tahiti, and elsewhere to return to England just as war was breaking out.[16] Mirroring the growth of travel aspirations amongst middle-class Britons through the end of the 1940s and into the 1950s,[17] the Features Department made recurrent series centred on producers' travels and their consequent impressions, often with titles formulated as *Portrait Of, Window On, Journey, or Return To or From.*[18]

MacNeice himself was no stranger to travel before he began his BBC career in 1941. He was already the author of two well-known travel books: with his friend the poet W. H. Auden, he wrote *Letters from Iceland* (1937), and a year later *I Crossed the Minch* (1938), about a Hebridean sojourn inspired in part by his Celtic roots, was published. He thus joined the rising numbers of interwar novelists and poets turning their hand to the travel book, including high modernists such as Ezra Pound, Wyndham Lewis, and E. E. Cummings. Travel signified a means to break with convention and tradition, but it also brought into view 'wrenching perspectives and uncanny displacements' that underlined modern alienation.[19] Visiting the Hebrides, wrote MacNeice, reminded him that he would 'always be an outsider' there as an Anglo-Irishman,[20] and yet London too would always remain to him 'a foreign city'.[21] The impulse to travel thus ran alongside, or emerged from, this sense of un-belonging and a desire to break down romanticised notions about unknown places and about travel itself.[22] A 'tripper's book' was the rather scathing term MacNeice used to describe *I Crossed the Minch* in its Introduction,[23] and, later, recalling his student days at Oxford, he punctured the pomposity with which he had once thought travel must be '"experience" at its highest'.[24] At the BBC, MacNeice found himself amongst colleagues who, like him, were drawn to travel and yet conscious of the 'uncanny displacements' it could produce.[25] In his features this generated a suspicion of travel's claims to authenticity and an aesthetics, to borrow Rebecca Walkowitz's terms, of 'estrangement' and 'hesitation'.[26] Walkowitz identifies such aesthetics as a form of cosmopolitanism that relocates modernism's internationalism away from traditional 'static models' of exile to a more expansive frame in which travel and mobility are key.[27] Extending the feature form into its second age,[28] MacNeice's vision of cosmopolitanism, as conjured up by travel, is one that is self-consciously hesitant and estranging not only with regards to the experience of journeying overseas but also in relation to the radio soundscapes that tried to bring it to life on air.

Jigsaws and Patchwork: Geography and Spatiality in the Radio Feature

MacNeice's first travel 'portrait' began life as a suggestion from an Italian BBC colleague, George R. Foa (later to become a well-known television director), who was sent to Rome to produce two of MacNeice's dramatic features for Radiotelevisione Italiana.[29] It was his first trip to the city and in initial discussions he referred to the programme as *Window on Rome*,[30] and it remains unclear when or why the title was changed before its broadcast on the Home Service on Sunday 22 June 1947 at 9.30 pm. The opening announcement, written by MacNeice to be read by the continuity announcer, may however offer some clues. Framing the feature as 'the merest adumbration', the announcement gives a sense of the broadcast ahead as a sonic portrait marked by a sense of its own construction.[31] Offering apologies to a long list of potential listeners, including scholars and lovers of Rome, MacNeice undercut BBC convention by playing down the authority of the production; *Portrait of Rome* was, he wrote, neither 'a lesson or a summary', but instead a 'patchwork' akin to a 'jigsaw with most of it missing'.[32]

Writing later in his poetry about the limits and possibilities of founding a 'castle on the air', as he put it in 'Autumn Sequel', his long poem of 1954,[33] MacNeice was demonstrably aware of the challenges involved in bringing an overseas landscape and culture to life through audio alone. The introductory strands to all three of his *Portraits* (including opening announcements, billings in the *Radio Times* and his own written introductions), stress their 'inadequate' and merely 'suggestive' attempts to build portraits of distant places against their geographical or 'physical background'.[34] A sense of distance and closeness, mediated through an emphasis on the geographical and the physical, runs right through MacNeice's travel portraits which are rich with details of streetscapes, buildings, windows and doorways and to broader, more topographical images, coalescing into a geographical and spatial—or geospatial—architecture that disorders the boundaries between 'here' and 'there', between the imagined landscape and the experienced soundscape. As the character of British expat Uncle Robin (played by Ronald Simpson) suggests in *Portrait of Rome* radio's reimaginings—its ability to shift location with few encumbrances in comparison to visual media—allow for a new kind of space–time matrix. Following a stop at the Sistine Chapel with his niece and her friend, who are tourists in the city, Uncle Robin suggests to the local tour guide that their view should be demolished and swapped for something free from the boundaries of time and space:

> And now let's take the chapel down again. Yes, and St. Peter's too. And half of these other churches and nearly all these palaces. Shove the streets closer together, let the balconies overhang more, run up a lot of brick towers anxious to butt one another. Step up the stink and the noise ...[35]

This radiogenic spectacle—one in which streets can be summoned up and be made to vanish in the space of a few seconds, mountains can be traversed, and the tourist, along with the listener, can ascend into the sky for a better vantage point—showcases a self-awareness about form and space. It illustrates Ian Whittington's assertion that MacNeice 'himself conceived of broadcasting as a medium of imagined spatiality'.[36] In self-consciously grappling with the aesthetics of spatial construction, *Portrait of Rome* gestures towards the wider evolution of radio grammar, when early to mid-twentieth-century radio forms wrestled with notions of spatial texture, representation of motion and audio perspective. For some there was a romance to the notion of radio as a limitless spatial imagination: as acclaimed BBC dramatist Donald McWhinnie recollected in his book *The Art of Radio* (1959), 'The radio performance ... makes possible a universe of shape, detail, emotion and idea which is bound by no limitations of space and capacity'.[37] For others, a cognisance of the technical and artistic limitations of radio-sound, and of the restrictions imposed by broadcast genres and networks, transmuted into programmes and studio design that bore the spatial hallmarks of their anxieties, as discussed by scholars including Neil Verma and Shundana Yusaf.[38]

The imbrication of space and geography in *Portrait of Rome*, and its two successor 'portrait' features, reveals MacNeice's exploration of space and form to be one marked above all by self-awareness. His portraits celebrate and complicate the idea of radio-spatiality, whilst also gesturing towards a knotty entanglement with overseas culture in the broadcast travel feature. As Home Service listeners heard, Uncle Robin's space-warping directions do little to soothe the tourists in *Portrait of Rome* who, like the travellers in the following two features, are full of bewilderment as travel brings in its wake cultural confusion and an acute sense of dislocation. Athens, for example, is a 'box on the ear, hard as nails',[39] the foreign locales are too hot and bothersome, and the hapless British and American tourists journeying through them want, periodically, to retreat. 'Athens gets me down', laments the nameless visitor in *Portrait of Athens*,[40] whilst the British protagonist in *Portrait of Delhi* is rather pointedly named 'Ignorance'.

There is a broader, dreamlike but unnerving quality to the visitors' experiences as spatial texture serves to both locate and to disorientate. No sound recording survives of *Portrait of Rome* but we can hear in the recording in the BBC's archive of *Portrait of Delhi* that the overall sound architecture is marked by a lack of signposting.[41] Voices intercut or emanate apparently from nowhere with little introduction and scenes move location and back and forth through time without conventional pauses and beats, often without the use of actuality to signal change.[42] 'Stop looking to the north/Look down',[43] says one of the guides amidst the confusion, attempting to scaffold the journey time and again through the use of repeated spatial directions. Yet when Ignorance does look down he sees a 'dreary landscape', full of 'flatness and monotony', prompting him to draw on the clichés of the foreign correspondent's vernacular, with the phrase 'Sorry, listeners' breaking the fourth wall.[44] Ignorance then declares the plains of northern India to be a 'featureless mottle of green and brown', whilst New Delhi is said to resemble a 'child's box of bricks'.[45] Radio-space can therefore expand or even disappear as characters play with the boundaries of their tourist personae, but as it intersects with the concrete geographical and cultural realities of a foreign place it can also become dull and devoid of dimensionality. In this way MacNeice's travel features showcase a very modernist engagement with the pull-and-push of space and place, through an exploration of the 'twin spatial visions' in which the former equates to movement and possibility, and the latter to a sense of stasis.[46]

To return to Uncle Robin and the mutability of land and soundscape in *Portrait of Rome,* it becomes apparent that in MacNeice's travelogues even firm ground can be elusive. The streets beneath the tourists' feet remain shaky, and the landscape sometimes literally shape-shifts making them further question their bearings. For Evelyn, Uncle Robin's niece, a journey to discover the history of Rome leads to being thrown right off balance:

Evelyn: What's wrong with the ground?
Uncle Robin: It's sinking.[47]

As the built environment of culture and history disappears, akin to 'dirty water when you let the plug out', the travellers sink further into the moulting streetscape until a crash landing unites them with the 'tessellated pavements' of third-century Rome.[48] This lack of firm ground, its porousness and mutability, situates MacNeice's radio aesthetics in a geomodernist frame,[49] and is significant in deciphering his interpretations of travel functioning not only as a metaphor for cross-cultural touristic interchange but also, by extension, for the foreign radio feature itself, which grapples with the slippery task of imprinting a lifelike geographical impression in the airwaves not through complete dramatic license nor through straight current affairs retelling.[50] Indeed, Uncle Robin notes:

> [H]istory is like ... geology in fact; you get all sorts of unconformities. Faults, as they say. What ought to be horizontal and buried away instead gets cooked up and comes to the surface, side by side with what's modern.[51]

So the very experience of modern, or modernist, travel is revealed as one that is muddled or 'cooked up', but so too is the 'patchwork' architecture of the radio travel feature, one that can only be as suggestive, to use MacNeice's words, as a jigsaw with most of it missing.

MacNeice may have been aware that some listeners, at least, may have expected a more straightforward approach to a foreign affairs feature, akin to 'a lesson or summary',[52] rather than a production style that, as it were, pulled the rug from under their feet. In a remarkable self-referential scene in *Portrait of Rome*, Evelyn and her friend Margaret discuss a radio programme with the very same name, anticipating the unfavourable reception these 'portraits' did go on to receive from many listeners:

EVELYN: Really! The BBC!
MARGARET: What is it, Evelyn dear? Where've you been?
EVELYN: Listening to the wireless ... .
MARGARET: Well, what was it all about?
EVELYN: Here. Rome. *Portrait of Rome* it was called, I only switched on because I was drying my hair—but really, darling, I'm not particularly stupid but I couldn't make out what they were getting at. One moment I thought it was Catholic propaganda and the next they were using bad language. And such a sort of hotch-potch!'.[53]

The 'hotch-potch' effect on the listenership was tangible. 'I think your *Portrait of Delhi* was drivel', wrote one Mrs Johnson directly to MacNeice after its transmission;[54] it was 'vulgar buffoonery' and a 'comic strip' according to Cambridge academic C. H. Dodd who was also moved to write in. Three years later, the Listener Research Report on *Portrait of Athens* gave it an Appreciation Index of 58, 'well below the Sunday night features average of 70', and lower than Laurence Gilliam's recently broadcast feature *Journey into Greece.*[55] Listeners to *Portrait of Athens* are said to have been critical of the 'ultra-modern impressionist standpoint' that resulted in an 'incoherent mass of local colour' and an 'over-produced, 'precious' programme'.[56] *The Manchester Guardian*'s review of *Portrait of Rome* similarly attacked the lack of a 'clear line' in a broadcast it felt was let down by MacNeice taking the word 'patchwork', it is said mistakenly, 'as a guiding star'.[57] Yet MacNeice's correspondence in response to personal letters addressing criticisms directly to him reveals an unwavering belief in the validity of features as a hybrid and avowedly modernist genre. Writing to one N. V. L Rybot he noted: 'A feature programme is not supposed to be like a talk—still less like the 9 o'clock news'.[58] Replying to C. H. Dodd he insisted, 'It strikes me that your poor opinion of the content of this programme must be due to a misunderstanding of its form', before citing modernist media techniques, namely montage, which were vital to his production artistry.[59] Overseas radio travel, as composed by MacNeice in the feature form, pushed geography and spatiality to the fore and in so doing transmuted into a bold radio aesthetic. Orientating listeners towards more complicated and experimental features, reliant in this case on the geospatial dimensions of radio 'abroad', was undoubtedly one clear aim of MacNeice's broadcasting work in the late 1940s and early 1950s. Yet another kind of reorientation was equally important—one that hinged on ideas of territory, conquest and boundaries between nations.

Birds, Planes and End of Empire

As part of the BBC's coverage of the momentous independence of India and its Partition in 1947, MacNeice was asked to gather material for a proposed six-part series, *India and Pakistan*, for the Features Department. Joining BBC colleagues Wynford Vaughan-Thomas, who was largely in a reporting role, and fellow Features producer Francis (Jack) Dillon, he arrived in Delhi only a few days ahead of the historic events of 14 and 15 August. Setting aside the copy of E. M. Forster's 1924 novel *A Passage to India* he had been reading in preparation for his trip, he reached instead of for 'an armful of newspapers' in an effort to bring himself up to speed with the latest political developments.[60] MacNeice's travels had not, as yet, taken him anywhere as far-flung, nor had the BBC tasked him with such a weighty responsibility before. He, along with his fellow producers, was to interpret for its Home Service and Third Programme listeners the first, pivotal moment in the dismantling of Britain's empire.

The letters he sent home from India to his wife Hedli Anderson began by relaying a sense of social whirl as the BBC staff were entertained by British officials and taken to meet leading Indian politicians. But as August drew to a close, the BBC team set out northwest from Delhi, aiming towards Peshawar, to gather features and news material from the field. Travelling via Punjab, they witnessed first-hand the communal violence and carnage of Partition, most notably in the town of Sheikhpura where they came across 1,500 injured Sikhs and Hindus taking refuge in a school building following a massacre, some 'with their hands cut off', smothered by 'hordes of flies' and with their clothes 'covered with rusty-brown dried blood'.[61] The incident jolted MacNeice from the pose of aloof onlooker and transformed him, according to Vaughan-Thomas, into a 'man of action' helping the refugees into lorries, ensuring their safe passage out of town, and firmly engaged as an outsider 'deeply and profoundly involved in the human dilemma'.[62] A few weeks later MacNeice referenced Punjab when writing to Laurence Gilliam with an impassioned plea for BBC Features to rethink its proposed coverage, and to tone down any celebration of the Raj's legacy so as to avoid connotations of gloating and back-slapping. He urged Gilliam to think carefully about the BBC's responsibility to be 'extra tactful', and to steer clear of the tone of many domestic newspapers that, MacNeice had heard, had 'seized on the Punjab and Delhi as a pretty occasion for an I-told-you-so vindication of the British Raj'. MacNeice urged that 'the BBC should on no account follow suit … To stress at this moment the security, justice, good order or unity bestowed by the Raj would be surely most inopportune'.[63]

Before sketching out his map for a revised series MacNeice delineated in strenuous terms the topics he felt should not be included, and these encompassed the army, democracy, religion and, above all, 'The Present Communal Trouble', which he argued would unfairly frame the subcontinent as a fractured nation bound in strife and animosity.[64] Instead, the proposal outlined in his letter accentuated the intertwining of geography and history so as to 'ram home to the listener from the start the peculiarities of India, physical and spiritual'.[65] Three of the six suggested features were to be loosely constructed around geographical tropes, including *Portrait of Delhi*, and Gilliam gave his consent for the series to go ahead. The reoriented Home Service productions that came to fruition in 1948 included Francis Dillon's *Indian Village* and *The Khyber Pass*, Nesta Pain's *'Lord, Look at my Field'*, MacNeice's *Portrait* and his final, summative examination in *The Road to Independence*: all aimed towards offering an expansive view 'of all the main elements of India'.[66]

Broadcast on the Home Service on Sunday 2 May 1948 (9.30–10.30 pm), *Portrait of Delhi* utilises the same geospatial mechanics as first encountered in *Portrait of Rome* a year earlier, but here the guides taking charge are personifications of India. Four figures appear as lead characters, and the first of these voices is 'Topography' (played by Susan Richards), followed by 'History' (Duncan McIntyre) and 'Literature' (Allan McClelland).[67] But the predominant guiding figure taking the character of Ignorance on his journey is a native bird (Reginald Beckwith). More specifically, it is a bird that periodically changes form, starting off as a carrion crow, turning into a bulbul and then a vulture because, as it asserts, 'my view depends on my mood—on which bird I am at the moment'.[68] A shape-shifting aerial trajectory affords Ignorance the opportunity to look down on a series of overlapping scenes ranging from key episodes in Hindu and Muslim history to the decisive juncture of the Great Mutiny of 1857 which locked Britain and India into a dichotomous and dialectical relationship, deepening faultlines that recast India and its populace as British subjects who must know their place as 'natives'.[69] The Bird laments how the earthbound restrictions of humans fortify boundaries:

> All those months of the siege the English were there on the Ridge and the sepoys were here on the walls – trying to get each other's range. And they may have got it in gunfire – but they never got it in spirit. Yet the birds could fly freely between ... [70]

The sky above, or the air, is thus configured as a place where cultural and national borders can be crossed and transgressed, and where transmogrification allows for divergent vantage points and perspectives. In this respect, the conceit of the Indian Bird typifies modernism's many engagements with notions of nation-space and aerial activity, echoing the use of aeroplanes in the novels of Virginia Woolf or the sky-views proffered by John Piper in his landscape paintings.[71]

Yet MacNeice's decision to shift the feature's topography airwards rather than staying on the ground as with *Portrait of Rome*, and later *Portrait of Athens*, was a curious choice given that India's unshackling from British rule, and the birth of Pakistan, was manufactured by a splicing—a very real partition—of the land of the subcontinent. The sidestepping of the very word 'partition' in *Portrait of Delhi* illustrates a sensitivity to the concept, also alluded to earlier in MacNeice's letter to Gilliam where he stated that a 'radio partitioning' of the features series would be an unwise editorial decision.[72]

Conceptions of geographical appropriation, spatial expansion and empire building would not, of course, have been alien to MacNeice as an Irishman. As Ashok Bery reveals, Irish echoes resounded loudly on the subcontinent for MacNeice, whose letters from South Asia often cast the two nations in a comparative light.[73] Writing home, he described the Pathans of the North West Frontier Province as 'v. Irish in their approach to things', and in another letter written while on a trip to Sri Lanka in 1955 he noted that the Indian Ocean on a grey day reminded him of Belfast.[74] Furthermore, in a 1950 essay on India for Laurence Gilliam's volume celebrating the work of the Features Department, he argued that 'educated Indians are as politics-ridden as the Irish', and went on to note, 'perhaps this is inevitable with subject or newly liberated peoples'.[75] Witnessing the partition of the subcontinent ignited memories for MacNeice, with all its attendant complexities, in what Rajeev Patke, in his study of Irish poets, terms 'partitions of the mind'.[76] This may illuminate why, when dealing explicitly for the first time with Britain's disintegrating empire in his radio work, MacNeice retreated from metaphors borne out of ground warfare, occupation, and fissure. Instead there is an embrace of the air (waves) and the space-binding nature of radio, of its links to notions of imagined community, and of its potential to carry a new kind of meaning—a transnational, cosmopolitan post-empire vision of the nation.

This is writ large in another piece of shape-shifting at the feature's end: it is not bombs, but steel birds, or planes, that shower Delhi's landscape with roses as Gandhi's funeral takes place. The Topography guide decides to silence herself, noting that 'Geography counts no longer; the world is one', and the Literature guide remarks 'Delhi is no longer a long way off'.[77] Hence, *Portrait of Delhi* registers the post-empire precipice on which Britain finds itself as India gains freedom, and in its orientation inwards and outwards: towards the centre of an imminently crumbling empire where Home Service listeners were clustered, and to a transformed global landscape where travel and redrawn national boundaries reveal glimpses of a world where nowhere is a long way off. In pushing his radio aesthetics, and the BBC, into the transnational sphere, into the liminal spaces between the end-of-empire-nation and the new states to which it gave birth, MacNeice offers a vision of airspace as home to border-crossing birds, planes and sound waves; as an interstitial territory where propaganda and spatial conquest give way to a reshaped global order and to a reimagined soundscape, and to a new geomodernism on the air.

Disclosure Statement

No potential conflict of interest was reported by the authors.

Notes

1. Coulton, *Louis MacNeice in the BBC*, 161.
2. *Countries in the Air* remained unfinished at the time of MacNeice's death in 1963 but two draft chapters were found amongst his papers at the BBC (letter from E.R. Dodds, MacNeice's literary executor, to Features Department, 9 November 1963, BBC WAC L1/285/1). These were published as one fragment, 'Landscapes of Childhood and Youth', in *The Strings Are False*, 216–38. For a synopsis of the book see Coulton, ibid., 161–162.

3. See the Appendix in Wrigley and Harrison for list of MacNeice's travel-related features.
4. As defined by Coulton, 149–173.
5. Doyle and Winkiel, *Geomodernisms*, 3–4.
6. One of the earliest such broadcasts was *Special Empire Day Programme* on 5XX Daventry (*Radio Times*, 24 May 1926, Issue 138, 333).
7. Aimed at schools and the general listener, *Travel Talk* was broadcast on 2LO London and 5XX Daventry from the end of the 1920s to the mid 1930s, after which it became a regular series on *The Children's Hour* on the Home Service.
8. Gielgud was a *Radio Times* editorial assistant in 1927, and was billed as a British writer with Polish connections for this programme. He held overall responsibility for radio drama production from 1929 until after the Second World War.
9. *Radio Times*, 2 September 1927, 22.
10. Hilmes, *Network Nations*, 2.
11. Ibid.
12. See also Bridson, *Prospero and Ariel*, 80.
13. 'Home Features', wrote Laurence Gilliam, was 'taking a new shape and intensity ... likely to be very much influenced by the growth of Overseas'. Memo to Val Gielgud, 4 February 1941, BBC WAC L2/78/1.
14. Heppenstall, *Portrait of the Artist*, 47–57, and Bridson, *Prospero and Ariel*, 133–150.
15. Gilliam, *BBC Features*. Scripts included *The End of Mussolini* by D. G. Bridson, *India at First Sight* by MacNeice and *From Anzio to Burgundy* by Wynford Vaughan Thomas.
16. Ibid., 97, 149.
17. Wakeman, "Veblen Redivivus," 427.
18. Examples include *Portrait of Istanbul*, 13 November 1949, and *Return to the U.S.A.*, 18 March 1952, both Home Service, prod. D. G. Bridson; *Return Journey: To Strasbourq*, 12 September 1948, Third Programme, prod. Rayner Heppenstall; *A Journey in Greece*, 6 November 1951, Home Service, prod. Laurence Gilliam.
19. Farley, *Modernist Travel Writing*, 1; Fussell, *Abroad*, 15–23.
20. MacNeice, *I Crossed the Minch*, 7.
21. MacNeice to T. S. Eliot, 25 February 1957, printed in Allison, *Letters*, 614–615.
22. Ibid. MacNeice writes that 'in foreign travel one is much of the time searching for the implementation of certain myths', but the insinuation is that often reality does not match travellers' expectations.
23. MacNeice, *I Crossed the Minch*, 7.
24. MacNeice, *The Strings are False*, 235.
25. See also Ranasinha, "South Asian Broadcasters in Britain," 2010, on the transcultural atmosphere in the BBC during wartime and exchanges between the BBC's Eastern Service staff, such as M. J. Tambimuttu, and writer-producers who also broadcast on the BBC's domestic services. MacNeice published work in *Poetry London*, a journal founded by Tambimuttu, and was included in Tambimuttu's edited anthology *Poetry in Wartime* (1942).
26. Walkowitz, *Cosmopolitan Style*, 16–20.
27. Ibid., 8. High modernist émigré authors included James Joyce, Gertrude Stein and T. S. Eliot.
28. I refer here to the post-war iteration of Features Department, following its separation from Drama Department in 1945, until its closure in 1964.

29. *The Dark Tower* and *The Golden Ass*. George R. Foa memo to Laurence Gilliam and others, 1 June 1947, BBC WAC R19/953.
30. Letter from MacNeice to George R. Foa, 31 March 1947, BBC WAC R19/953.
31. Opening Announcement, Programme Report, *Portrait of Rome*, ibid.
32. Ibid.
33. MacNeice, "Autumn Sequel," *Collected Poems*, 387.
34. Amended billing, *Portrait of Rome,* 4 June,1947, BBC WAC R19/953.
35. *Portrait of Rome* script, 20, line 197, BBC WAC.
36. Whittington, "Archaeologies of Sound," 48.
37. McWhinnie, *The Art of Radio*, 37.
38. Verma, *Theater of the Mind*, 33–72; Yusaf, *Broadcasting Buildings*, 20–27.
39. *Portrait of Athens*, 18 November 1951. Time code: 00'30–00'34. Recording accessible in the British Library Sound Archive (BLSA). Script accessible in the Harry Ransom Center, University of Texas at Austin.
40. Ibid., time code: 17'06–17'08.
41. A lack of signposting is notable when compared to earlier features about place, even those produced by MacNeice himself such as *The Stones Cry Out series* (1941, BBC Home Service).
42. Ibid., time code: 08'16; *Portrait of Delhi* recording also in BLSA, time code: 01'31.
43. *Portrait of Delhi* script, 7, line 77. BBC WAC.
44. Ibid., line 75.
45. Ibid., line 76.
46. Thacker, *Moving Through Modernity*, 13.
47. *Portrait of Rome* script, 23, line 224–5.
48. Ibid.
49. Doyle and Winkiel, Introduction.
50. See Introduction to this special issue for discussion of features as a genre.
51. *Portrait of Rome* script, 16, line 158.
52. Ibid., 23, lines 225–226.
53. Ibid., 15, lines 147–151. See also '*What Now?*', part of MacNeice's series *The Four Freedoms*, in which two radio listeners, one of whom is called Evelyn, discuss whether they can detect propaganda in a radio feature (BBC Home Service, 28 March 1943, 8.45 pm).
54. Letter from Mrs Johnson to MacNeice, 2 May 1948, BBC WAC R19/948.
55. *Journey Into Greece* scored 69% in the Index; Listener Research Report, *Portrait of Athens*, BBC WAC LR/51/2556, Week 47. On MacNeice and other writers' radio engagements with Greece, ancient and modern, see Wrigley, *Greece on Air.*
56. Ibid.
57. *The Manchester Guardian*, 25 June 1947, 3.
58. BBC WAC R19/R53.
59. Letter from MacNeice to C. H. Dodd, 13 May 1948, BBC WAC R19/948.
60. Stallworthy, *Louis MacNeice*, 354–366.
61. Quotations from MacNeice's letter to Hedli MacNeice, 31 August 1947, in Allison, *Letters,* 487. For fuller description of the trip, see 470–511.
62. As quoted in Allison, 488, n. 1.
63. Letter from MacNeice to Laurence Gilliam, 19 September 1947, in Allison, 496–502.
64. Ibid., 498.

65. Ibid.
66. Ibid. The series was broadcast on Sundays from 9.30 pm to 10.30 pm between 18 April and 15 August 1948, BBC Home Service. MacNeice also produced *India at First Sight* for BBC Third Programme (13 March 1948, 8.30 pm).
67. *Radio Times*, 30 April 1948, 8.
68. *Portrait of Delhi* script, 2, line 18. Perhaps MacNeice drew inspiration from Forster's use of a native bird as metonym for India. See *A Passage to India*, 1924, Chapter 8.
69. Ibid., 40, line 457.
70. Ibid., 40, line 455.
71. Beer, "The Island and the Aeroplane," 265–290; Harris, *Romantic Moderns*, 25–29. See also how MacNeice explores this idea in the passage of his 1946 *Enemy of Cant* which draws on Aristophanes' *Birds*, as discussed in Wrigley and Harrison, *Louis MacNeice*, 255–260.
72. Letter, 19 September 1947, in Allison, 498.
73. Bery, *Cultural Translation*, 74–100.
74. Letters to Hedli Anderson, 5 September 1947 and 5 November 1955, in Allison, 489, 598.
75. Gilliam, *BBC Features*, 61.
76. Patke, "Partition," 5, 17.
77. *Portrait of Delhi* script, 52, lines 569, 570.

Bibliography

Allison, Jonathan, ed. *Letters of Louis MacNeice*. London: Faber and Faber, 2010.

Beer, Gillian. "The Island and the Aeroplane." In *Nation and Narration*, edited by Homi Bhabha, 265–290. London: Routledge, 1990.

Bery, Ashok. *Cultural Translation and Postcolonial Poetry*. New York: Palgrave Macmillan, 2007.

Bridson, D. G. *Prospero and Ariel: The Rise and Fall of Radio – A Personal Recollection*. London: Gollancz, 1971.

Coulton, Barbara. *Louis MacNeice in the BBC*. London: Faber and Faber, 1980.

Doyle, Laura, and Winkiel Laura, ed. *Geomodernisms: Race, Language, Modernity*. Bloomington: Indiana University Press, 2005.

Farley, D. G. *Modernist Travel Writing: Intellectuals Abroad*. Columbia, MO: University of Missouri Press, 2010.

Fussell, Paul. *Abroad: British Literary Traveling Between the Wars*. New York: Oxford University Press, 1980.

Gilliam, Laurence, ed. *BBC Features*. London: Evans Brothers in arrangement with the British Broadcasting Corporation, 1950.

Harris, Alexandra. *Romantic Moderns: English Writers, Artists and the Imagination from Virginia Woolf to John Piper*. London: Thames and Hudson, 2010.

Heppenstall, Rayner. *Portrait of the Artist as a Professional Man*. London: Peter Owen, 1969.

Hilmes, Michele. *Network Nations: A Transnational History of British and American Broadcasting*. New York: Routledge, 2011.

MacNeice, Louis. *The Strings are False: An Unfinished Autobiography*. London: Faber and Faber, 1996 (first published 1965).

MacNeice, Louis. *Collected Poems*. London: Faber and Faber, 2007.

MacNeice, Louis. *I Crossed the Minch*. Edinburgh: Polygon, 2007 (first published 1938).

McWhinnie, Donald. *The Art of Radio*. London: Faber and Faber, 1959.

Patke, Rajeev. "Partition and Its Aftermath: Poetry and History in Northern Ireland." *Journal of Postcolonial Writing* 46, no. 1 (2010): 17–30.

Ranasinha, Ruvani. "South Asian Broadcasters in Britain and the BBC: Talking to India (1941–1943)." *South Asian Diaspora* 2, no. 1 (March 2010): 57–71.

Stallworthy, Jon. *Louis MacNeice: A Biography*. London: Faber and Faber, 1995.

Thacker, Andrew. *Moving through Modernity: Space and Geography in Modernism*. Manchester: Manchester University Press, 2003.

Verma, Neil. *Theater of the Mind: Imagination, Aesthetics, and American Radio Drama*. Chicago: University of Chicago Press, 2012.

Wakeman, Rosemary. "'Veblen Redivivus': Leisure and Excess in Europe." In *The Oxford Handbook of Postwar European History*, edited by Dan Stone, 423–442. Oxford: Oxford University Press, 2012.

Walkowitz, Rebecca. *Cosmopolitan Style: Modernism Beyond the Nation*. New York: Columbia University Press, 2007.

Whitehead, Kate. *The Third Programme: A Literary History*. Oxford: Clarendon Press, 1989.

Whittington, Ian. "Archaeologies of Sound: Reconstructing Louis MacNeice's Wartime Radio Publics." *Modernist Cultures* 10, no. 1 (Spring 2015): 44–61.

Wrigley, Amanda. *Greece on Air: Engagements with Ancient Greece on BBC Radio, 1920s–1960s*. Oxford: Oxford University Press, 2015.

Wrigley, Amanda, and S. J. Harrison, eds. *Louis MacNeice: The Classical Radio Plays*. Oxford: Oxford University Press, 2013.

Yusaf, Shundana. *Broadcasting Buildings: Architecture on the Wireless 1927–1945*. Cambridge, MA: MIT Press, 2014.

WHO'S LISTENING TO MODERNISM? BBC FEATURES AND AUDIENCE RESPONSE

Alexandra Lawrie

*This article focuses on three literary features—*A Room of One's Own *(1946), centred on the work of Virginia Woolf, an adaptation of David Jones' epic poem* In Parenthesis *(1946) and Herbert Read's verse play* Moon's Farm *(1951)—in order to examine how they reflected on and adapted various aspects of modernist innovation. It also examines the range of listener response offered in the BBC Listener Research Reports for these programmes—from surprise and delight to outright bewilderment, with many listeners pleased in principle with the challenging literary content, but looking for pointed guidance on how to approach it, whilst others who were receptive to avant-garde writing were quick to identify lapses in quality. This article therefore traces how literary modernism functioned, and was experienced, when adapted into an aural experience in the late 1940s and early '50s, the most innovative period of the Features Department.*

The BBC Features Department was *the* space for experimentation in broadcasting from the mid-1930s until its closure in 1964, attracting many of the period's most talented writers and producers including Elizabeth Bowen, Rayner Heppenstall, Louis MacNeice, Seán Ó'Faoláin and Dylan Thomas, and led by the enormously committed Head of Features, Laurence Gilliam. Kate Whitehead has identified that in the first decade after the Prix Italia competition was established in 1948, practically all of the BBC's entries for the award were selected from Features, and all of its winners.[1] The Features output was rich and varied, incorporating programmes on science, travel, geography, literature, history and biography, particularly during the 1940s and '50s, when the Department was producing its most innovative work. The focus of this article is on a sample of the literary features it produced and, in particular, the ways in which these programmes reflected on and adapted various aspects of modernist innovation, as well as examining how listeners responded to these features. It will therefore begin by offering some detail about the reputation of the Department at that time. Thereafter it will describe the work of the BBC's Listener Research Section, which was established in 1936 to gather information about the public's response to programmes in general. It will then concentrate on three features from the late 1940s and early '50s, each of which were either focused on the work of or written by an important modernist writer—Virginia Woolf, David Jones, and Herbert Read—examining the content of these programmes, the modernist techniques and innovations they adopted and the impact of these broadcasts on listeners in order to trace the effects of literary modernism when adapted into an aural experience.

The feature format was a notoriously difficult one to define. In Gilliam's introduction to his 1950 book on the subject, he posed the question, 'What then, precisely, is the feature programme?', before admitting, 'There is no precise answer'.[2] Because, while the plays produced by the BBC Drama Department (headed by Val Gielgud) derived from the conventions of theatre, features were a fresh, hybrid form developed specifically for radio that borrowed from documentary, drama, and the radio talk, without exactly emulating any of these forms. This presented the Department with much more freedom: Douglas Cleverdon, one of its most lauded producers, described the feature form as having 'no rules', and delighted in the scope for experimentation this allowed.[3]

The Listener Research Section was set up in the mid-1930s for a variety of reasons, not the least of which was an effort on the part of the BBC to appear more responsive: given that the Corporation was funded by public money, it was prudent to be seen to seek public opinion on what this money was being spent on. Although a Programme Correspondence Section had been established in 1924 to respond to the more than 50,000 letters received by the BBC every year by 1927,[4] those who wrote into the Corporation were of course self-selecting individuals whose views could not be taken as a fully representative sample of the general public, and a more systematic method of gathering information was therefore mooted—even though this initiative was itself imperfect. But it is worth noting at this point that not all the pressure for audience research was coming from outside the Corporation; many producers also felt they were 'broadcasting into a void', with no clear sense of how their programmes were being received by ordinary listeners.[5] The BBC therefore hired Robert Silvey, a statistician for the London press exchange, to formulate various ways of sampling public opinion on a selection of programmes. It was made clear from the outset, however, that Audience Research had a limited remit: it was for information purposes only, and unlikely to have a direct impact on policy or scheduling. Although Silvey created various different methods of measuring and sampling public opinion on programmes (for instance the Variety Listening Barometer, and the Continuous Survey of Listening), my interest lies with the Features Panel, set up in 1942 and comprising of listeners who tuned in regularly to features and were willing to provide feedback on a volunteer basis. This panel had a fluctuating membership which was maintained at around the 600 mark, with an average length of service of 18 months. A confidential BBC report of 1944 stated that every effort was made to ensure that the structure of the panel corresponded to the feature programmes public more generally; their comments, therefore, can be taken as broadly representative. Many of these listeners, the report also revealed, used features for educational purposes, with programmes on scientific discoveries, travel, history, and biography in highest demand. But it is also worth acknowledging at this stage that Audience Research was not without its detractors. In Silvey's account of his time at the helm he describes how some of his new colleagues at the BBC 'dismissed market research as so much ballyhoo ... the Board of Governors, they argued, were the representatives not the delegates of the public and its decision to embark on audience research amounted to an abdication of its responsibilities'.[6]

Panellists' duties were simply to complete questionnaires (which they were sent) about the features they listened to, with their answers providing the raw material for the reaction reports produced by Silvey's department. These reports usually ran to one or one-and-a-half pages per programme, consisting of an 'Appreciation Index' (the mark out of a ten given by each panellist, averaged out to a percentage), an estimate of general audience size, an overview of the response to the programme, and some sample quotations from a handful of respondents. Each report was circulated to the relevant producer, the head of their department, and the schedulers, although anyone in the BBC was permitted to see a copy. This was not, of course, a perfect model for gathering audience feedback: a larger panel would have provided a more accurate picture of how each programme was received; the fact that not all programmes were surveyed means that feedback cannot be compared across a department's entire output; and the reports that were produced for individual programmes were far from extensive. But the reports do, nevertheless, provide us with a constructive if partial impression of how the programmes examined in this article were regarded at the time of broadcast.

A Room of One's Own (1946, Home Service)

The remainder of this article will focus on three features which were either centred on the work of or written by an author whose work has come to be associated with literary modernism, and tracing the audience response to each. The first of these is *A Room of One's Own*, a 30-minute feature about Virginia Woolf and her work, written by the novelist John Hampson, produced by Edward Livesey and broadcast on the Home Service at 7.00 pm on Sunday 25 August 1946.[7] Programmes of this type functioned as didactic tools, but the tone was also inclusive and encouraging, motivating listeners to broaden their cultural horizons after hearing them. This accorded with the aims of the first Director-General John Reith, who had left the BBC in 1938: although he was certainly no admirer of modernist literature, in more general terms he had been determined to use the wireless as a way of fostering cultural tastes. Thereafter, during William Haley's tenure from 1944 to 1952, the model of a 'cultural pyramid' was mooted—the idea being that by providing listeners with small amounts of more serious material on the Light Programme, they would feel encouraged to move up to the mixed programming provided on the Home Service, and eventually then graduate to the Third Programme, the highest cultural echelon, which was established in 1946. The feature on Virginia Woolf, a Home Service broadcast, worked in this fashion by taking a 'difficult' modernist writer and successfully encouraging listeners to start reading her work. But while the author and material under discussion was undoubtedly challenging, the programme itself was not presented in a particularly innovative or complex format; it was conversational, informing, and reasonably light-hearted.

According to the research report, this programme on Woolf commanded only a small audience—around 2% of the listening public—but it was noted that a study of the life of poet William Cowper, *Deity Disowns Me*, which was broadcast a week earlier in the same slot, had an audience of a similar size, suggesting that it may have been the timing, rather than the subject matter, that kept the audience figure low.[8] The format of the feature is of a conversation between two men—a Man and a Business Man. But there is a third character, too—the popular radio actress Marjorie Westbury plays the role of Virginia Woolf, and she cuts in with lines or passages from Woolf's writing to illustrate the case the Man is putting to the Business Man, which is ultimately to convince the Business Man, a reluctant reader, of the quality of Woolf's work. The Man is a recent convert to Woolf, and is trying to persuade the Business Man—and, by extension, the listener—to start reading her.

This Home Service broadcast went out a month before the BBC's dedicated 'high cultural' Third Programme was launched, and was therefore aimed at a broad listenership, but the script testifies that it did not shy away from discussing some of the innovative aspects of Woolf's writing. A line from *Mrs Dalloway*—'Life, London: this moment of June'—becomes the keynote of the broadcast, and fundamental to how her work is understood. At every juncture Woolf's affection for the city streets, with the immediacy of 'everyday life' is emphasised. But it is made clear that Woolf also looks deeper than other writers, reaching for the material they might easily overlook, in order to communicate life's fluidity: 'She rescues the treasures which we conceal under the mound of common experience. She rifles dumps and rubbish heaps to produce jewels and precious fragments' (15). And the Man continues, in a rather dreamy fashion, to describe what he perceives to be the effect of Woolf's prose:

> In reading the novels of Virginia Woolf you will find again and again that like a poet she catches a fleeting moment or mood, which by the delicacy of her perception she fixes exactly in all its exquisite freshness and beauty so that it endures. (17)

The Business Man, playing the role of gruff philistine, swiftly dismisses this perceived loftiness, objecting to the lack of facts, or supposed real life, in her novels: 'Why doesn't she tell us how rich people are, how much land they own, how many bedrooms they have and what kind of car they keep. Something solid. Any child can blow bubbles, or look through coloured glass'. At this point Westbury steps in with a passage from the end of *To the Lighthouse*, when Lily Briscoe, having returned to her painting after a lapse of ten years, thinks back to Mrs Ramsay, and recognises her ability to take ordinary moments from everyday life and shape them into something articulate and emphatic—into something akin to art:

WOOLF: What is the meaning of life? That was all. A simple question, one that tended to close in on one with years. The great revelation had never come. The great revelation perhaps never did come. Instead there were little daily miracles, illuminations, matches struck unexpectedly in the dark, here was one. This, that, and the other, herself and Charles Tansley, and the breaking wave; Mrs Ramsay bringing them together; Mrs Ramsay saying 'Life stand still here'; Mrs Ramsay making of the moment something permanent. (18)

By quoting this passage, the point is made, however subtly, that Woolf's great achievement was to produce art out of the quotidian, and that listeners might reasonably turn to her novels as the means to cast new light on familiar aspects of their own lives. Reading Woolf, it is suggested, will encourage them to see things in new and alternative ways. This serves as a rejoinder to the Business Man's complaint that Woolf's writing seemed without purpose, offering nothing either 'solid' or 'serious'.

Woolf's own personality is also discussed. She is described by the Man as a 'Lover of life' who 'once played a part in a great hoax' (10), disguising herself as an Ethiopian prince and smuggling aboard a battleship, and also accepting a bet to write a book that would be chosen as a Book of the Month club selection:

B.M. What an absurd story. Why she didn't know a thing about popular appeal.

MAN: Didn't she? Anyway she knew all about the English passion for dogs. And her book *Flush*, the biography of Elizabeth Barratt-Browning's spaniel had an enormous public success. And what is more to the point, it was the first choice of a Book Society. And that is true, even if the anecdote is false. The legends are numerous – but never believe the one which tells you that she turned her back on life. (10–11)

This focus on the more light-hearted aspects of Woolf's personality was key: the writer of the script John Hampson never shied away from the fact that 'She is not an easy writer', but he was also keen to foreground her more human aspects, because this would render her more approachable for the ordinary, interested reader, and thereby encourage them to try her novels.

The research report demonstrates that this feature chimed with listeners: the quotations from the novels Westbury read out were felt to have been 'not only delightful in themselves, but also representative of the writer's works', while the 'questions and answers' between the two main characters—one enthusiastic, the other sceptical—was a 'great help to better understanding of the writer', addressing the same types of reservations that the some listeners had also felt. But this feature did more than simply offer entertainment and interest over a half-hour period: based on the sample of quotations in the report, the programme also stimulated some listeners who had previously felt intimidated by Woolf's reputation to try reading some of her work for themselves. One respondent, described as a bank inspector, offered the following comment:

> Until now, I have always understood Virginia Woolf to be overwhelmingly highbrow and unreadable. The broadcast perhaps put matters into truer perspective – it certainly interested me, and I feel that I must make her closer acquaintance – even though a little warily.

Another, an art teacher, felt encouraged to make another attempt at reading her work:

> I had only struggled with one of Virginia Woolf's books, and had never reached the end – now, having heard this programme, I feel I would like to have another try. Some bits were lovely, and much easier to understand than I had imagined [she] could ever be.

Not all of the feedback was positive. Some 'manifestly had no interest in Virginia Woolf', found the programme 'vague, highbrow or boring', or were 'unable to understand her'. But if we are to see these quotations as representative of a more general response (and the careful composition of the listener panel suggests we should do just that), there is still plenty of evidence that large numbers of listeners used literary features as a 'way in' to approach the more 'difficult', modernist authors.

In Parenthesis (1946, Third Programme)

By contrast, radio adaptations of *In Parenthesis*, David Jones' epic poem of the First World War published in 1937, were modernist in presentation, style and content. Douglas Cleverdon eagerly took on the task of adapting it for radio presentation: it had been due for broadcast on Saturday 11 November 1939 but with reduced programming following the recent outbreak of war his production was shelved. Cleverdon's production was announced in *Radio Times* for broadcast on Tuesday 10 November 1942 but this too was abandoned at the last moment in favour of a broadcast by Winston Churchill. Cleverdon's adaptation was not, in fact, broadcast until 1946, when it went out on the Third Programme on Tuesday 19 November (8.30–10.00 pm), preceded by an introduction by David Jones himself. This 1946 production received regular repeat broadcasts into early 1947. On Thursday 11 November 1948 (8.25–10.00 pm), a 'New production of the broadcast on November 19, 1946' (with lead roles taken by the same actors as in 1946) went out for Armistice Day, again with Cleverdon producing; as indeed he did once more for the a brand new, and longer, production which was aired on Sunday 30 January 1955 (3.00–5.00 pm).[9]

Jones' long prose poem—acclaimed by W. H. Auden, W. B. Yeats, and most famously T. S. Eliot, whose influence helped get it published by Faber and whose 'Note of Introduction' is printed in most modern editions—is widely regarded as a significant modernist work: certainly Eliot argued that it was as important as the later work of James Joyce and the *Cantos* of Ezra Pound.[10] It is also one of the very few which dealt directly with the events of the First World War. The poem employs what is recognised as the 'mythical method' Eliot described in '*Ulysses*, Order, and Myth', and implemented in 'The Waste Land'—drawing on the mythic past as a way of giving order and shape to contemporary history.[11] Jones utilises the early medieval Welsh epic *Y Gododdin*, the Bible, texts by Shakespeare and Sir Thomas Malory, and others dating from ancient Rome onwards, as paradigms or templates for conveying, in a seven-part narrative form, his own six-month experience as a soldier in the trenches between December 1915 and July 1916. Almost all of the soldiers encountered in the poem will die, except for the central character John Ball (played by Emrys Jones in 1946 and 1948 and Richard Burton in 1955; Stephen Murray had been due to take the role in 1942). Fred D. Crawford suggests that Jones manages to position the First World War 'as part of a tradition' but 'without ennobling the struggle'.[12] He does this by indicating the presence, or survival, of past conflicts: the voices of soldiers from ancient and medieval battles percolate the narrative, and are conflated with those of Welsh and English soldiers in the 1916 battalion. There is at all times the sense of a deeper parallel world mirroring that of the present day: John Ball's experiences of trench life are rendered on a mundane level, with the tedium of routine tasks and conversations with fellow soldiers, but at any moment the narrative might shift in time and space to accommodate the alternative visions or experiences Ball is able to call forth. This might be knights seeking the Holy Grail, ancient Greece as described by Herodotus or soldiers on their way to Agincourt.

The production of Jones' work for radio in 1946 presented Cleverdon with significant challenges, not least the difficulty of cutting the poem into a length suitable for a single 90-minute broadcast. By looking at his 1946 script and cross-referencing this with the poem, it is possible to trace how the editing was carried out. As noted above, the original version contained unsignalled shifts between unnamed voices speaking in fragmentary dialogue, interlaced with literary allusions and digressions into an ancient Celtic underworld. Cleverdon overcame some of these difficulties of comprehension by creating three central voices to act as narrators: the Narrative of Action, with the most lines, which set the scene and described movements and events; the Narrative of Thought, which conveyed the emotions of the soldiers (most commonly fear and anxiety) in a poetic, contemplative tone; and the Narrative of Memory, with the fewest lines, usually referring back to Malory, or occasionally another folk tale. Aside from these three voices, the script also consisted of conversations taking place between the soldiers, but here rendered much more comprehensible than in the original poem. In order to demonstrate the latter point, it is worth comparing a section from the original poem with the abridged version from the broadcast script. In Part 7 of Jones' original poem, tension is high as the soldiers are counting down towards battle:

> But they already look at their watches and it is zero minus seven minutes.
> Seven minutes to go ... and seventy times seven times to the minute.
> this drumming of the diaphragm.
> From deeply inward thumping all through you beating
> no peace to be still in
> and no one is there not anyone to stop
> can't anyone – someone turn off the tap
> or won't any one before it snaps.
> Racked out to another turn of the screw
> The acceleration heightens;
> The sensibility of these instruments to register,
> fails;
> needle dithers disorientate.
> The responsive mercury plays laggard to such fevers – you simply can't take any more in. And the surfeit of fear steadies to dumb incognition, so that when they give the order to move upward to align with 'A', hugged already just under the lip of the acclivity inches below where his traversing machine-guns perforate to powder white [...]
> you have not capacity for added fear only the limbs are leaden
> to negotiate the slope and rifles all out of balance ... [13]

Cleverdon's 1946 script reveals the kind of changes he made:

JENKINS: Zero minus seven minutes, Sergeant.
SERGEANT: Seven minutes to go ... .
NARR. OF THOUGHT: And seventy times seven times to the minute, this drumming of the diaphragm.
NARR. OF ACTION: Racked out to another turn of the screw the acceleration heightens.
EFFECT: (MACHINE GUN FIRE)
NARR. OF THOUGHT: The surfeit of fear steadies to dumb incognition.
BALL: When they give the order to move upward to align with A, you have not the capacity for added fear, only the limbs are leaden to negotiate the slope. (48–9)

This example shows that large chunks of unassigned text are broken up by Cleverdon, with lines given to narrators and soldiers, and amended to work as dramatic scenes. Listeners appreciated his work on the script: 70% of respondents praised it in the highest terms, commenting on the 'vivid picture' created, the 'glowing imagery of the language', and the 'interweaving of thought and action'.[14] These comments provide us with a valuable insight into how high modernism—albeit an adapted, abridged and simplified version—was being received over the wireless. Cleverdon's creation of three distinct narrative voices, set apart from the soldiers' dialogue, was also regarded as a particular triumph, as the following remark suggests: 'It was vivid, poetic and arresting, with fine contrast between the private's educated and philosophical thoughts on the one hand and his immediate 'earthy' existence on the other'. As one may expect with such a challenging piece of writing (and, indeed, with any kind of broadcast to a mass audience), the response was not one of unanimous approval: the research report also noted that

> small groups of listeners found the method of presentation too confused for their understanding, too broken by different kinds of narration and overcharged by esoteric thought sequences which inevitably reduced the speed of the action. Even some of the listeners who admired the script found that it required a sustained effort of concentration which was tiring.

But this type of comment was not altogether surprising—the feature, though simplified, was not intended to be simplistic, and some of the shifts in tone are difficult to follow. This is particularly the case when the feature presents sections of soldiers' dialogue, immediately followed by the Narrative of Memory, the allusive, folk-tale narrator:

LEWIS: Sights at 350.
BALL: Yes, Corporal.
LEWIS: Starving as brass monkeys. Diawl!
N. OF MEMORY: Starved as Pen Nant Govid, on the confines of hell, where sits that wintry hag, the black sorceress. (17)

Yet without this stark and potentially alienating juxtaposition between these two forms of voice—one pragmatic and colloquial, and the other lofty and mystical—much of the force and character of the original poem would have been lost. Although the effect is jarring, we should recall that in the original, there is no line or voice distinction whatsoever, so it is for the individual reader to deduce who may be speaking or thinking any of those lines, rendered as follows: 'Starving as brass monkeys—as the Arctic bear's arse—Diawl!—starved as Pen Nant Govid, on the confines of hell'.[15]

Rose Macaulay reviewed this feature for her 'Critic on the Air' slot on the Third Programme. There she suggested that Cleverdon had actually improved on the original version of the poem, noting that 'for the first time it was clear who said which things', and commending Cleverdon for noticing the 'improbability' of a soldier's remark being followed, 'as if made by the same man, by a phrase or sentence out of character—a literary allusion, a quotation from Mallory [sic] or Milton or the Bible, or merely an over poetic piece of prose', and instead opting to give these literary quotations to one of the three narrators 'who take charge of the commentary when it isn't actually uttered by speakers'.[16] Her ringing endorsement for this feature can be seen as representative of the enormously positive press response to the programme, but we might be wary about praise that appears to applaud deviations from the poem's original modernist form. Cleverdon made the changes he did in order to render it suitable for broadcast. He needed to set the scene, create parts for different actors, and to integrate those esoteric allusions in a manageable way.

Macaulay gives the impression that Cleverdon had found something lacking in the original, and set out to rewrite it in a more realist style. But in fact he appears to have got the balance right: the mechanics of broadcasting demanded that he build in different characters, yet he still managed to retain the philosophical planes and linguistic textures that render Jones' poem so distinctive.

The listeners, more so than Macaulay, seem to have recognised this: a housewife was quoted in the report as saying of the actors' voices, 'They were as threads to make a harmonious tapestry of one portion of life as lived then and in that place'.[17] This was remarkably close to Jones' own conception of the work: as he explained in the Preface (which he was also heard reading out at the start of the 1946 production), the poem was designed to be an experiment in writing, 'using as data the complex of sights, sounds, fears, hopes, apprehensions, smells, things exterior and interior, the landscape and paraphernalia of that singular time and of those particular men'.[18]

Moon's Farm (1951, Third Programme)

In Parenthesis was a very different type of feature to the one on Virginia Woolf: it was dense and dramatic rather than explanatory and conversational, demanding far higher levels of commitment from the listener.[19] That it managed to elicit a largely favourable audience reaction demonstrates how Third Programme listeners were willing to engage with difficult modernist literature in a feature format. But not all Third Programme feature programmes hit the mark and negative receptions are revealing of both tastes and expectations.

The 1951 programme *Moon's Farm*, written by Herbert Read and produced by his friend and fellow Yorkshireman Rayner Heppenstall for broadcast on Sunday 21 January (18.55-pm) as part of the series 'The Inward Eye', was a poetic rendering of Read moving back to a part of Yorkshire where he had spent his childhood.[20] The work was later be published as part of a longer collection of poems in 1955, and it belongs to Read's late period. It is a verse play written for three voices, with the overall theme as one of return. The Second Voice has come back to the site of the eponymous now-vanished farm, where he encounters two others: the First Voice, played in the BBC production by a woman, representing 'place'; and the Third Voice, played by a man, representing 'time'.

Moon's Farm opens with the First Voice describing her current homelessness, and looking back to an earlier time when Moon's Farm was being built:

> A cave in the old quarry, a dry ditch, a broken-down barn – that is all my shelter these days. The sheep have gone, and with them the shepherds. That was twenty winters ago, and then came the men with axes and cut down the spinneys and plantations, lopped off the branches and carried the timber away, and left this desolation at the head of the Dale. (1)

This elegiac tone is retained as the play develops; at certain junctures an Aeolian harp is played, further contributing to the ethereal, numinous character of the piece. But also noticeable throughout, however, are the modernist elements. Read first met T. S. Eliot in 1917, and they became close friends, with Read contributing to *The Criterion* on a regular basis. By the 1930s Read was better known as an art critic, but he always saw himself as a poet first, and some of his early volumes in particular were written in an imagist style. Although his modernist tendencies were tempered in later years by a poetic philosophy which was increasingly indebted to the pastoral modes and settings of Romantic poetry, his facility for economic precision, and for crystallising a striking image, betray his longstanding literary connection to the work of Eliot and Pound. *Moon's Farm* is concerned with memory and a return to the land, but formally and also in terms of word choice, the poem still feels recognisably modernist: the three voices are palpably disembodied, with little attempt to make them into rounded characters. The reader (or in this case, and in the first instance, the listener) is also struck by a series of concise but fragmented images, and the recourse to myth as a means to emphasise the diminished contemporary circumstances:

> The Greeks were right again: it is life that plays game of hide-and-seek – the rhythm of the seasons as an interplay of life and death. In the person of Persephone it is life itself that disappears for a wintry season. But hiding involves a hiding-place. In Eleusis you can still see the pit by which Persephone went down into Pluto's dark realm. All the ancient myths are precisely located. We have no myths today because we have no sense of place – the beliefs of our age are like gas-filled balloons, untethered. They disappear into the clouds – into the transcendental inane. (21)

The audience research report for the 1951 production of *Moon's Farm* indicates that its appreciation index was 59, close to the average of 58 for previous programmes in 'The Inward Eye' series, although below the current average of 65 for Third Programme features in general.[21] Response to its content was far from laudatory: slightly more than a quarter of the panel 'found it 'a little obscure' but were 'set thinking' by it', but many 'dismissed it as 'a disjointed intellectual ramble', bearing no relation to its subject'. The individual comments from respondents which were included on the report (presumably because they were deemed broadly representative) allow us to drill down for a more detailed view. One described *Moon's Farm* as 'ideal broadcasting material', 'A sensitive and moving account up to Herbert Read's best standard'; another commended its 'clarity of vision'. But the negative responses are more intriguing, both in terms of their specific objections to the programme, and also the fact that they were offered up by individuals—a writer, a student, a university lecturer, and a journalist—whom one might reasonably expect to be more sympathetic to Read's experimental enterprise. The first of these, for instance, the 'Writer', answered as follows:

> What can one say of this flimsy conversation piece? Its method of dialogue – 'he speaks, then I speak, then he speaks again,' etc. – left an impression of pedestrianism that wasn't entirely justified by the contents of those speeches. It doesn't seem to achieve anything. I believe the substance would have been better delivered as a verbal essay.

The chief objection here was to the perceived failed connection between the three voices, which were felt to be little more than disengaged mouthpieces for conveying insubstantial ideas. The 'Journalist' respondent, meanwhile, disliked Read's allusions to other literary texts throughout the piece:

> Would have liked to hear pure Read, uninterrupted by King Lear and Persephone who annoyed and distracted me. One just got absorbed in his theme about the 'Logos' etc. and then Lear butted in, or Persephone to say his bed was made, or the pee-wit had a few words to say. Most irritatingly disrupted and it had such immeasurable potentialities.

And finally the university lecturer suggested that 'A little more local colour would have built up the atmosphere more successfully ... Not enough was done to help the listener to share some of the author's feeling for the place'. This last reservation was shared by the wife of a factory inspector:

> I did not see what purpose was served by the "genius loci" and her aged father, of whose identity I was never sure. The prose was good to listen to, however, and a something emerged, but for me it was not connected in the least with a sense of place.

What is most striking in these comments is the impatience with those aspects of the play we would identify as recognisably modernist: those disembodied voices, the oblique references to other literary works, and the discomfiting sense of dislocation, or unreality, in terms of setting. Another comment, from the wife of a metal turner, sums up this general confusion:

> It isn't often I listen to the Third Programme. The title, *Moon's Farm*, drew me, but the play, well, it must have been too highbrow like chamber music. Usually I can remember a play and tell my husband all about it, but this one, nothing at all, it was queer.

Given the markedly more enthusiastic response accorded to *In Parenthesis*, we can infer that modernist features (and perhaps features in general) were most successful when they were introduced thoroughly, giving the listener a chance to approach the work from a place of familiarity: as mentioned above, the radio adaptation of *In Parenthesis* included a long preface from David Jones in which he explained the wartime context of the piece, and prepared listeners for the mixture of Welsh and English voices, dialogue and liturgy. *Moon's Farm*, lacking this type of framing, left many listeners simply baffled.

Conclusion

Clearly the response to *Moon's Farm* contrasts with the more enthusiastic reports on the Virginia Woolf programme and the adaptation of *In Parenthesis*. Clearly, when the material being covered was perceived to be challenging, listeners responded well to gentle guidance on how to approach the work and what to look out for—although it is worth pointing out that in the feature on Woolf, her most experimental works, such as *Jacob's Room* and *The Waves*, were barely mentioned at all, as if the techniques on display in these novels might prove too alienating, or simply could not be explained within a half-hour programme for the pre-Third Programme Home Service audience. As noted above, *Moon's Farm* was part of the series 'The Inward Eye'; these types of series, which were premised on a particular theme or idea, were created as a way of encouraging authors to produce new material for the BBC. The wireless feature was still to some extent an unfamiliar format, and one within which many writers were unwilling to work, so the logic went that if the central premise was already there (in this case, describing a subject or events into which the writer had a particular insight, and couching it in a poetic narrative), the writer only needed to find their own take on it. But many of these series, 'The Inward Eye' among them, still struggled to find writers willing to produce material, and quality could sometimes suffer as producers such as Heppenstall and Cleverdon scouted around for contributors, and the series itself was shifted to different producers. The radio adaptation of *In Parenthesis*, on the other hand, had been the 'long-standing proposal' of Cleverdon, and the production featured 'quasi-liturgical interludes' written by the renowned composer Elizabeth Poston and sung by Henry Washington's Schola Polyphonica choir,[22] while major acting roles were, across different productions, taken by Leonard Sachs, Dylan Thomas, and Richard Burton. In general, therefore, this was simply a better-quality production, based on source material that was indisputably first rate. By the early 1950s listeners had become discerning arbiters of feature programmes, responding to them, on the whole, as culturally beneficial moments, and when they failed to deliver, they were criticised and producers and writers urged to do much better.

Disclosure Statement

No potential conflict of interest was reported by the author.

Notes

1. Whitehead, *Third Programme*, 119.
2. Gilliam, *BBC Features*, 9.
3. Cleverdon, *Growth of Milk Wood*, 17.
4. Briggs, *Golden Age of Wireless*, 67.
5. Whitehead, *Third Programme*, 53.
6. Silvey, *Who's Listening?*, 32.
7. Page numbers in brackets in this section refer to the script which can be consulted in the BBC Written Archives Centre.
8. Listener Research Report for *A Room of One's Own*, 5 September 1946, LR/6073, BBC Written Archives Centre.
9. Page numbers in brackets in this section refer to the 1946 script which can be consulted in the BBC Written Archives Centre.
10. **Eliot reference to follow.
11. **Another Eliot reference to follow.
12. Crawford, *British Poets*, 228.
13. Jones, *In Parenthesis*, 155–6.
14. Listener Research Report for *In Parenthesis*, 2 December 1946, LR/6740, BBC Written Archives Centre.
15. Jones, *In Parenthesis*, 52.
16. Macaulay, "Critic on the Air," 1.

17. Listener Research Report for *In Parenthesis*.
18. Jones, *In Parenthesis*, x.
19. Listener Research Report for *In Parenthesis*.
20. Page numbers in brackets in this section refer to the 1951 script which can be consulted in the BBC Written Archives Centre.
21. Listener Research Report for *Moon's Farm*, 13 February 1951, LR/51/181, BBC Written Archives Centre.
22. Cleverdon, "David Jones and Broadcasting," 73.

Bibliography

Briggs, Asa. *The Golden Age of Wireless*. London: Oxford University Press, 1965.

Cleverdon, Douglas. *The Growth of Milk Wood*. London: J. M. Dent, 1969.

Cleverdon, Douglas. "David Jones and Broadcasting." *Poetry Wales* 8, no. 3 (1972): 72–81.

Crawford, Fred D. *British Poets of the Great War*. London: Associated University Presses, 1988.

Gilliam, Laurence, ed. *BBC Features*. London: Evans Brothers, 1950.

Jones, David. *In Parenthesis*. London: Faber and Faber, 2010.

Macaulay, Rose. "The Critic on the Air." Third Programme Broadcast, 21 Nov. 1946. BBC Written Archives Centre.

Silvey, Robert. *Who's Listening? The Story of BBC Audience Research*. Birkenhead: George Allen & Unwin, 1974.

Whitehead, Kate. *The Third Programme: A Literary History*. Oxford: Clarendon, 1989.

INTERMEDIAL RELATIONSHIPS OF RADIO FEATURES WITH DENIS MITCHELL'S AND PHILIP DONNELLAN'S EARLY TELEVISION DOCUMENTARIES

E. Charlotte Stevens and **John Wyver**

*Writing of the closure in early 1965 of the Radio Features Department, Asa Briggs identifies one of the reasons for the controversial decision as 'the incursion of television, which was developing its own features.' '[Laurence] Gilliam and his closest colleagues believed in the unique merits of "pure radio". The screen seemed a barrier' (*The History of Broadcasting in the United Kingdom*, Vol. 5, p. 348). Rather than the screen being 'a barrier' for them, a number of the creators of the emerging television documentary were from the late 1950s onwards able to transfer and transform distinctive techniques of 'pure radio' into highly effective visual forms. Two key figures were the producers of 'poetic' documentaries Denis Mitchell and Philip Donnellan, who employed layered voices, imaginative deployments of music and effects, and allusive juxtapositions of sound and image, to develop an alternative (although always marginal) tradition to the supposedly objective approaches of current affairs and, later,* verité *filmmakers. And a dozen years after the dismemberment of the Features Department, Donnellan paid tribute to it in his glorious but little-seen film* Pure Radio *(BBC1, 3 November 1977). Taking important early films by Mitchell and Donnellan as case studies, this paper explores the impact of radio features on television documentaries in the 1950s and early 1960s, and assesses the extent to which the screen in its intermedial relationships with 'pure radio' was a barrier or, in the work of certain creators, an augmentation.*

The output of BBC Radio's Features department is widely acknowledged as one of the glories of post-war broadcasting. Many of the department's programmes are held to exemplify 'the one unique form that radio had achieved in its short history ... vital, contemporary, experimental, and above all "pure radio"'.[1] These transmissions were, as Virginia Madsen has described, 'somewhere between drama, documentary, poetry, and what the BBC also called "panoramas"—grand, large-scale productions based on fact, but rendered dramatically ... the best that the BBC had to offer.'[2] Radio Features closed in early 1965, and one of the reasons for the decision is identified by Asa Briggs as 'the incursion of television, which was developing its own features. [Laurence] Gilliam and his closest colleagues believed in the unique merits of "pure radio". The screen seemed a barrier.'[3] Television documentaries by the mid-1960s were doing much that had been the province of Features, but it is in the work of two filmmakers in particular that certain of the distinctive techniques of 'pure radio' were extended into highly effective visual forms.

The early films of Denis Mitchell, including *Night in the City* (1957) and *Morning in the Streets* (1959), and of Philip Donnellan, of which *Joe the Chainsmith* (1958) and *Coventry Kids* (1960) are characteristic examples, employed layered voices, imaginative blends of music and audio effects, and allusive juxtapositions of sound and image to develop a self-consciously 'poetic' form of small-screen documentary. They also integrated techniques more usually associated with filmed drama within documentary frameworks, such as the staging of fictional scenes, albeit with their real-life protagonists. Moreover, Mitchell and Donnellan continued certain of the political concerns of the output of radio features, highlighting regional characters and concerns from beyond the metropolitan focus of much other television and giving prominence to the marginalised voices of working people. This article explores the ways in which Mitchell and Donnellan formed their distinctive styles and suggests how their films contributed to an alternative, and subsequently marginalised, strand within British television documentary alongside the supposedly objective, narrative-driven, journalist-led approaches of current affairs and, later, the forms of *verité* filmmakers.

Documentaries made in the late 1950s and '60s by Mitchell and Donnellan can be seen to extend elements of a British documentary tradition identifiable in films from the 1930s and '40s by Basil Wright, such as *The Song of Ceylon* (1934), and by Humphrey Jennings, including *Listen to Britain* (1942, co-directed by Stewart McAllister) and *A Diary for Timothy* (1945). They complement too aspects of the documentaries associated with the polemical stance of the 'Free Cinema' group in the 1950s by aspirant directors including Lindsay Anderson, Karel Reisz and Tony Richardson. This alternative tradition can also be recognised later in aspects of the work of filmmakers including Robert Vas, who worked for BBC Television from the mid-1960s to 1978, and Mike Dibb, whose BBC films include *Seeing Through Drawing* (1976) and *The Country and the City* (1979). In the 1970s this alternative tradition of authored and allusive filmmaking found a sympathetic home in the Music and Arts department of BBC Television. It was for that department's strand *Omnibus* that, a dozen years after the dismemberment of Features, Donnellan paid tribute to its radio broadcasts in his documentary *Pure Radio* (1977). Elegantly filmed in monochrome (with only an introductory section in colour), the film combined interviews with some of those who created Features programmes, including Charles Parker, with a reconstruction of a radio studio in rehearsal. Similarly recognising the significance of Features, this article draws out the connections and parallels between that department's output and the early television work of Mitchell and Donnellan, whilst also contributing to the exploration of the intermedial relationships between radio and television at a formative moment. In the late 1950s the younger medium was for the first time reaching a mass audience at the same time as, and in part as a consequence of, the challenge to the BBC's monopoly posed by the arrival in late 1955 of the commercial network ITV.

First Features and Early Documentaries

'Before the war,' Paddy Scannell has noted, 'the development of the radio feature was uneven and chequered.'[4] A 'feature', *Radio Times* explained as early as 1928, was a special programme demonstrating 'original forms of expression, peculiar to radio,' and among the first such broadcasts were the innovative and impressionistic radio essays produced by Lance Sieveking in the late 1920s.[5] In the following decade, the London-based Talks department encouraged strands of reportage that engaged with the social and political questions of the day. Producer Laurence Gilliam also began to experiment with actuality sound recordings made on location, but this technique was exploited most effectively in programmes made in Manchester, where it was combined with an imperative to bring to the airwaves the life and language of northern England. The poet and producer D.G. Bridson oversaw a sequence of scripted feature programmes about local industries and explored new ways of working with verse and music, and from 1937 Olive Shapley used a new Mobile Recording Unit to document people speaking in their homes, clubs and workplaces. Both Bridson and Shapley foregrounded the voices of working men and women, although recorded actuality would not become central to radio's output until portable magnetic tape recorders began to be widely used in the mid-1950s. Meanwhile, in London, feature programmes were beginning to develop a more literary focus, with writers like Rayner Heppenstall and John Pudney becoming producers, a trend that was to become even more marked during World War Two with the arrival of Louis MacNeice as a writer-producer and the involvement also of, among others, John Betjeman, Dylan Thomas and Geoffrey Grigson. As Asa Briggs has observed, 'There was an intimate connection between feature-writing and poetry.'[6] Having flourished during the war, Features became a separate department in 1945 under Gilliam who was forthright in proselytising for 'the one unique form that radio has achieved in its short history. Owing something to the radio play, something to the radio talk, it is a synthesis different in essence from either.'[7]

By the mid-1950s, BBC Television had yet to develop factual programming that was comparable in achievement or distinction to the radio features from Manchester in the 1930s or from Gilliam's post-war Features department. In television, programmes of this type were more usually identified as documentaries, although they shared little with the dominant film documentary tradition established in Britain by John Grierson and others in the 1930s.[8] In 1955, in a book about how to write for television, Arthur Swinson identified three forms of television documentary: the dramatised documentary, scripted and mounted with actors 'live' in a studio; the magazine documentary, showcasing intriguing characters and travelogue segments; and the actuality documentary, which like the dramatised documentary combined studio and filmed elements, but with a stronger sense of contemporary journalism.[9] The most prominent example of the actuality documentary was the series *Special Enquiry*, which began in 1952 under editor Norman Swallow.[10] Despite's Grierson's former colleague Paul Rotha having been appointed as television's Head of Documentary Programmes in 1953,[11] the service had made only a small number of factual films, in part because of the high cost of working with 35 mm film stock but also because of opposition from the cinema industry and its unions. This would change, however, as television began to adopt the more affordable, though lower-resolution, 16 mm format in the later 1950s for 'news and current affairs, as these areas were regarded as more journalistic than aesthetic'.[12] Denis Mitchell and Philip Donnellan would be among the first generation of filmmakers to take up the creative challenges offered by these institutional and technological changes.

Denis Mitchell: Features for Radio, Films for Television

The BBC Television film *Night in the City* opens with a high-angle shot of a woman alone walking away from the camera down a flight of steps.[13] Her face is hidden, her step lacks urgency and her heels clip loudly on the stone, all of which suggests that she may be a prostitute. As she turns a corner to continue her descent, a graphic of the film's title is overlaid and a plaintive guitar theme breaks in. Setting, subject and mood are all suggestive of *film noir*, as is the title itself, which echoes that of crime movies like *Cry of the City* (Robert Siodmak, 1948) and *Night and the City* (Jules Dassin, 1950). The *noir*-ish tone and alluring sense of mystery continue as a male narrator speaks: 'At night the familiar shabby city is changed by silence and by shadow into an unknown land. It's this unknown land I want to explore ... ' Two years earlier, a radio feature also called *Night in the City* was broadcast by BBC Home Service North.[14] This begins with a woman's voice reflecting on the sadness of walking the city streets after dark before the male narrator breaks in: 'Past midnight. Time to explore the city. This symbol of art and order. This beehive. *Explore*, for it's no longer workaday and familiar; it's an unknown land.'[15] Both film and radio feature were produced by the BBC North Region in Manchester, both featured music composed by Ewan MacColl, and both were devised, produced and narrated by Denis Mitchell.

The son of a Congregationalist minister, Mitchell had moved to South Africa at the age of 18. After the war he was employed by the South Africa Broadcasting Corporation and became fascinated by the possibilities of working with wire-recorders left behind by American troops. By the time BBC producer D.G. Bridson arrived in South Africa in 1949 on a year's secondment Mitchell was convinced that 'the ability to record people talking at their jobs and in their homes was not a mere novelty but a most important means of communication.'[16] Bridson was impressed with Mitchell and persuaded him to return to Britain and to work in the Radio Features department. After a short time there he joined BBC North Region where he further developed his techniques of recording ordinary people's voices on location and presenting them without over-insistent mediation or commentary. 'Mitchell experimented with a portable recorder,' Philip Purser noted in a 1961 profile of the producer, 'taking it into the streets and pubs and revivalist chapels of the North and bringing back hour upon hour of random talk. Edited, shaped and presented as a regular feature, *People Talking*, it lent a flavour and spontaneity to the spoken word undreamt of by Lord Reith.'[17] Norman Swallow similarly recognised how significant was Mitchell's use of the portable quarter-inch tape recorder that became available in the mid-1950s, noting his 'use of this new equipment for his own personal essays in radio-verité had already generated much excitement in broadcasting circles, who now brought it to television.'[18]

After a television training attachment at Lime Grove in 1955, Mitchell refined his technique by complementing his voices with images. The television documentary *Night in the City* followed earlier films about teenagers and prisoners, and while it makes use of a narrator, in many ways it employs classic techniques associated with the Features department. Bridson recognised this when he paid tribute to Mitchell's early films:

> The only [television films] that appealed to me were the finely observed documentary studies of Northern life being written and directed by Denis Mitchell. But as these were

> generally frowned on by his colleagues as being 'more radio than television' there seemed little point in ploughing a similar furrow.[19]

What is immediately apparent from the opening of the television version of *Night in the City* is that this is a film in which sound will play a leading, and perhaps even dominant, role. 'Documentary' elements are mixed with dramatised sequences, and the anonymous voices of ordinary people are central. Music beyond MacColl's score is also fundamental to the film, with sequences of teenagers dancing to rock'n'roll, nuns singing their night office at the Poor Clares convent in Levenshulme, and a faith-healing service at the Sharon Full Gospel Church in Stretford. Apart from a notional progression through a single night, *Night in the City* has no over-arching narrative, and instead this is an evocative mood piece which two years after transmission the filmmaker Karel Reisz described as possessing 'an astonishing feeling of unity, a strong pervading sense of a single sensibility responding to wildly diverse material.'[20] Reisz lays stress on Mitchell's singular authorship of the film, reflecting the importance of personal expression that was so prized by Gilliam and his colleagues. Gilliam had argued that:

> The essential quality of the feature programme is that it should be the expression of one mind, whatever technique it uses. It is the answer that each individual writer finds to the problem of making a statement by broadcasting, with the greatest possible force and coherence, emotional and dramatic impact, best suited to the nature of his material.[21]

The focus on authorship also chimes with the concerns of Karel Reisz and others, including the emerging directors Tony Richardson and Lindsay Anderson, that were demonstrated by the documentaries being promoted under the 'Free Cinema' banner. In the late 1950s these distinctive, independently-produced films, each of which stressed a personal, authorial vision, attracted significant attention as alternatives to the dominant documentary tradition. Reisz and Richardson's *Momma Don't Allow*, for example, which was first shown in the initial Free Cinema programme at the National Film Theatre in February 1956 is a vibrant yet intimate study of a Saturday night in a north London jazz club. Its respectful and affectionate, portrait of young people at play has a tone close to that of Mitchell's study. Reisz and Richardson's film also shares with *Night in the City* what Christophe Dupin has identified in many Free Cinema films as 'a creative way in which the filmmakers arranged sounds (often a combination of natural sounds and added music) and images, often creating symbolic contrasts between them, the rhythm of the films existing only as an amalgam of the two.'[22]

The visuals in *Night in the City* were shot with cumbersome 35 mm cameras, which in part accounts for the artificial, 'studio' quality of certain sequences. For the later, Prix Italia-winning television documentary *Morning in the Streets*, however, Mitchell and his camera operator collaborators Roy Harris (who also co-directed), Gerry Pullen, Graham Turner and Ted Wallbank, used lightweight 16 mm cameras new to the BBC Northern Film Unit.[23] These cameras are usually associated with their flexibility in newsgathering and with *verité* filmmaking pioneered by Jean Rouch and Edgar Morin in France and by Richard Leacock in the United States, but their poetic potential was also considerable, as *Morning in the Streets* demonstrates. But like the earlier film, *Morning in the Streets* was also based on a radio documentary, *The Talking Streets* (1958), arranged and produced by Mitchell, and described in the *Radio Times* billing as 'An impression of life and opinion in the back streets of a Northern city in the morning.'

The film begins with panoramic shots of a generic city (scenes were shot in Manchester, Salford, Stockport and Liverpool) before focussing in on streets of working-class housing, populated as dawn breaks only by pigeons and cats. A family—two adults and five children sleeping in just two beds—wakes to face the day. Scenes of home life are juxtaposed with the street games of children, of housewives at the shops, of road building, of house construction, and of unemployed men standing around and taking refuge in a library. Throughout we hear voices speaking about the harshness of life, expressing dreams of the countryside, exchanging gossip, much of the time accompanied by an under-score composed by Thomas Henderson with Tommy Reilly on solo harmonica. The dense weave of words from voices that remain for the most part unidentified ranges across living standards, class identity, social aspirations, memories of the war and connections with global events, especially the threat of the atom bomb, even if to a viewer today it is striking that there is no explicit mention of immigration or racial tensions (although a male voice says, in a friendly manner, ' ... there's that many Africans over here').

The film combines its allusive impressionism with sequences that are more directly narrativised. We meet a philosophical vagrant evading the police, and Mitchell constructs montage juxtapositions of images in which, for example, a motorbike being kick-started is immediately followed by a reclining doll apparently animated into a seated position. There are also sections with synchronous sound (where the person speaking is also shown on screen), as when (in an echo of Arthur Elton and Edgar Anstey's 1935 documentary *Housing Problems*) a tenant addresses the camera to describe the terrible conditions of their rented accommodation. The most clearly constructed sequence draws together a radio broadcast of the Remembrance ceremony at the Cenotaph with a widow's wartime reminiscence of her husband leaving home cut with shots of fallen toy soldiers left over from a boy's playtime. Towards its close, the film concentrates more strongly on the activities, faces, and voices of children, as John Corner has recognised:

> Over a selection of shots of individual girls, smiling and singing in their 'solo' and 'chorus' roles, their song is suddenly taken up orchestrally. As it increases in volume, the deserted playground featured in the early morning sequence is suddenly filled by a rush of boys and girls. Underneath the music and in relation to its volume and buoyancy, the camera now picks them out and frames them individually in a mode not of observed particular action but of projected general significance. Their faces stand for the future.[24]

Morning in the Streets is a rich trace of the failed post-war consensus and a striking riposte to 'You've never had it so good' Tory boosterism. In its rejection of narration, its weave of voices, its contrapuntal juxtaposition of image and sound, its exploration of a theme rather than adherence to a narrative, the film can be seen as aspiring to and achieving a form of 'pure television' aligned with the 'pure radio' of the Features department. In the three decades after *Morning in the Streets*, Denis Mitchell directed and produced a remarkable range of documentaries, shot both in Britain and abroad, many of which continued to develop these formal approaches. In British television during these years perhaps the only figure with a comparably rich and varied oeuvre, and with a filmmaking voice as distinctive, was Philip Donnellan, who similarly extended the 'pure radio' concerns of Radio Features to the small screen.

The Television Documentaries of Philip Donnellan

Philip Donnellan joined the BBC in 1948, and soon started writing and producing his own radio documentaries in Birmingham, before becoming Charles Parker's deputy in the Features department. In his unpublished memoirs, Donnellan writes (with characteristic directness) that BBC Television's Outside Broadcast Unit's output in the early 1950s was characterised by 'stiff and deferential' interviews and an overall air of 'unimaginable banality'.[25] Inspired by Free Cinema and other new international documentary film work, he came to believe that the captivating potential of television was not being adequately exploited, and he was certain he had the necessary skills and experience to make innovative films for broadcast. While Donnellan was equally dismissive of his previous radio work,[26] in his time with Features he became adept in working with magnetic tape, and in drawing out working-class interview subjects from around the Black Country and wider Midlands.

Donnellan was interested in the potential for authorial control in film documentary production 'just as one did with a simple programme on sound tape'.[27] Following a location-based secondment with Ealing Studios,[28] Donnellan attended a Gilliam-organised conference for Features producers, where Grace Wyndham Goldie (then Assistant Head of Talks, Television) spoke about writing for television: Donnellan writes 'she told of the potential wonders of studio-based "features"' for television, and he pursued an introduction.[29] Two days later, Donnellan met with Wyndham Goldie and producer Donald Baverstock, who were interested in what 'regional knowledge' Donnellan might bring to the second series of Baverstock's 'pioneering magazine programme' *Tonight* (1957–65).[30]

Donnellan's previous work with Features provided a model of poetic documentary-making as well as local stories to draw on: he produced diverting segments for *Tonight* featuring Midlands tortoise races and 'exotic' Birmingham junk shops, and in return gained experience in articulating television's 'visual grammar'.[31] Out of this work, Donnellan was permitted to contribute a film to *Eye to Eye*, the series for which *Night in the City* had been produced. Permission came after he appealed for parity across the regions: 'Denis Mitchell from Manchester was contributing a film; why not a film from Birmingham?'[32] The result was *Joe the Chainsmith* (1958).

Donnellan was very aware of Mitchell and his innovative radio work, particularly his use of the then-novel 'midget' tape recorder, writing that:

> Mitchell grasped its possibilities at once: he took his machine into the clubs, the stews, the lodging houses of Manchester, and with scarcely a word, holding the mic with one hand and a ciggie with the other, encouraging people only with his look of battered and opaque world-weariness, he recorded 'People Talking' [tx. 1953–8]. It was remarkable. A new Mayhew. A combination sometimes of Dostoievsky and Jack London ...[33]

Such ideas would have been in the background as Donnellan collaborated with another Birmingham producer Phil Drabble (later the presenter of *One Man and his Dog*) on a Midland Home Service radio series called *Men, Women and Memories* (1954–5), and it was on this that Donnellan journeyed into the enclosed industrial community of Cradley Heath near Birmingham, right at the heart of the Black Country, and where he first met chain-maker Joe Mallen. *Joe the Chainsmith* was Donnellan's first film that put into practice the principles of poetic television.[34]

Donnellan points out that *Joe the Chainsmith* is indebted more to his experience in radio rather than an awareness of television production, confessing that he 'had very few preconceptions about the way in which one might actually make a full-length film: in the same way, presumably, as one would make a radio programme', albeit with a visual dimension.[35] One especially striking sequence in the film, a poetic portrait of Mallen's daily life at work and leisure, is an extended observation of Mallen and colleagues hammering a ring of steel. The effect of the images and the sounds is hypnotic and transforms the sequence into, in the words of Franklin, 'a ritual of aesthetic wonder.'[36] Much of the documentary is shot away from the workplace: when Mallen shares advice given to him by his father ('a rough diamond ... a typical Black Country man'), he is filmed looking directly into camera, and we can see and hear the yard—complete with chickens and terrier—in which he sits. Mallen is presented as part of this milieu, enmeshed in a way of life seemingly untouched by modernity. By interviewing Mallen in the yard, the film illustrates, as Drabble noted, a way of life that even then were 'relics of an age that is passing'.[37]

Franklin remarked upon the film's focus on leisure pursuits, since in contrast to other contemporary representations, it 'represents the working class at play, rather than as proletariat'.[38] This is achieved through allowing these activities to be narrated by Mallen, and their association through video and audio montage of the Black Country village, working practices, and voices. Donnellan argued that Mallen's narration was a radical choice, since at the time, 'only the elect in broadcasting—commentators, interviewers, mediators—[were] allowed to speak facing the audience directly'.[39] Contrary to the original script, Donnellan and Drabble elected that Joe would 'do his own explaining' direct to camera, just as one might record an interview on tape.[40] Adopting production practices from Features offered a way of making television that could overcome what Donnellan saw as the BBC's class-bound limitations governing the ways in which, as Paul Long has written, 'it spoke of and to its audience' by allowing subjects to speak for themselves.[41]

In ways comparable to Mitchell's early films, techniques and concerns associated with pure radio are brought into play: voices from below, and in this case the surprising employment of Mallen himself as the host and dominant voice of the programme; the use of evocative music; impressionistic images employed allusively and associatively; and a concern to capture atmosphere and the textures of everyday life and leisure. For Franklin, Denis Mitchell's work and the early films of Donnellan are drawn together by their shared origin in the regions, a 'poetic and disjunctive aesthetic and humanist structure of feeling',[42] a contrapuntal relationship between sound and image grounded in the use of the tape recorder, and a central interest in *people*—as opposed to the social processes and institutional operations that so fascinated many of the films made within the documentary tradition associated with John Grierson.

Donnellan's *Coventry Kids* (1960) offers a compellingly ambivalent glimpse at Coventry's post-war resurrection. It features an innovative score by Ewan MacColl and Peggy Seeger, with the voices of Coventry residents presented over images of the city's wartime ruins, rebuilding efforts, industry, and diverse residents. Whether this film exhibits 'respect for people's individuality'[43] is a matter of debate. Many voices heard on the soundtrack for *Coventry Kids* are not identified, making the particular recorded statements into a generalised Coventry voice, even as the accents heard range from the local to the international.[44] In comparison to a mode of documentary filmmaking that would have a single narrator speaking for Coventry's residents, *Coventry Kids* does evoke a respect for the power of an individual voice to become a signifier of more universal experience.

The intermedial extension of the aesthetics of Features into this poetic mode of television documentary is most directly realised in Donnellan's television adaptations of certain of the innovative Radio Ballads. Produced between 1958 and 1964 by Charles Parker, Ewan MacColl (who had scored Mitchell's *Night in the City*) and Peggy Seeger, these radio programmes weave together the vernacularly-phrased and regionally-accented voices of ordinary Britons at work with specially-composed folksongs and ballads. Donnellan saw in the success of the Radio Ballads a 'permanent model of voice-music montage and therefore the stuff of documentary film', in both technical and political accomplishments.[45] Undeterred by a few false starts in realising this intermedial extension of radio to television,[46] the opportunity soon arose (following a change in BBC2 management) for another attempt.[47] Donnellan adapted three Radio Ballads successfully: *Singing the Fishing* (1960), which became *The Shoals of Herring* (1972); *The Fight Game* (BBC Home Service, 1963; BBC2, 1973); and *The Big Hewer* (BBC Home Service, 1961; BBC2, 1974). Alun Howkins has argued that these films 'bore the unmistakable stamp of the [Radio Ballad] form' in their translation to television.[48] Rather than illustrating the existing audio, Donnellan updated and expanded the scope of the original programmes: for example, with *The Shoals of Herring,* instead of cutting archive footage into an 'antiquarian and elegiac' film, Donnellan travelled to Scotland in 1972 to document fishermen's contemporary concerns.[49] These often-political additions provoked the ire of BBC brass,[50] as did his decision to screen (without adequate permission) *The Shoals of Herring* for those Scottish fishing communities, who could not receive BBC2.[51]

Donnellan's later films are held by some critics to have become too partisan, too 'political', although they long remained one of the few powerful expressions of progressive and radical ideas within BBC Television. His *Gone for a Soldier* (1980), combines witness statement, archive, and popular song in a critique of 150 years of imperial aggression, up to and including the 'politically sensitive' context of Northern Ireland.[52] But his film *Pure Radio*, broadcast on 3 November 1977, in the *Omnibus* series was less contentious. Donnellan's retrospective view in his memoirs argued not just for the film's importance as a memorial for the people and practices of Features, and as a 'culmination of our work in Television on the Radio Ballad form', but as a challenge against pressures for change within and beyond the BBC:

> it was also intended as a warning against those forces already on the move, indeed already largely dominant, which would curb such qualities in broadcasting in favour of a narrower hegemony: a sharp and shallow professionalism controlled by accountants and answerable to nothing. We were moving into a period, I reckoned, when the talents which Features had represented would in a dozen years be quite forgotten if we had not already built their tribute into the cultural archive of television.[53]

The irony is that *Pure Radio* was only broadcast once and forty years after its transmission archival copies of Donnellan's attempt at memorialisation are not easily accessible for viewing.

Conclusion

In the years preceding the closure of Features, the work of the filmmakers Denis Mitchell and Phillip Donnellan demonstrated that the television screen was not, as Asa Briggs suggested, 'a barrier' to translating the ideas of pure radio to the newer medium. Mitchell and Donnellan used the emerging technologies of 16 mm film cameras and portable quarter-inch tape recorders to extend techniques developed with radio feature programmes into poetic forms of television documentaries. Building on the richly developed techniques of Features, they worked with allusive combinations of spoken word and music with images, and they evolved self-consciously 'poetic' filmic forms. They extended the concerns of radio features to engage with stories from beyond London and to bring to the screen the marginalised voices of workers and of families living in poverty. Both Mitchell and Donnellan continued to direct films into the 1980s and beyond, and although their concerns, techniques and driving ideas certainly never became dominant in factual programmes, comparable approaches, and indeed traces of the ambition of 'pure radio' can be recognised in the films of later directors including Robert Vas and Mike Dibb. Yet if the films of Mitchell and Donnellan have to date received only minimal critical engagement, to which this article has made a modest contribution, the programmes to which Vas, Dibb and others contributed so creatively remain, both for filmmakers today and for commentators, almost entirely unknown and undocumented.

Disclosure Statement

No potential conflict of interest was reported by the authors.

Notes

1. Thomas, "Gilliam, Laurence Duval (1907–1964)," no page.
2. Madsen, "Radio and the Documentary Imagination," 192.
3. Briggs, *History of Broadcasting: Competition*, 348.
4. Scannell, "The Stuff of Radio," 23.
5. "Feature Programmes," *Radio Times*, 367; for Sieveking, see Scannell, "The Stuff of Radio," 1–7; also Hendy, "Painting with Sound," 169–200.
6. Briggs, *History of Broadcasting: Sound and Vision*, 702.
7. Gilliam, "Aspects of the Feature Programme," 100.
8. Key books about the Grierson tradition include Aitken, *Film and Reform*; Anthony and Mansell, *Projection of Britain*; Sussex, *Rise and Fall*; Swann, *British Documentary Film Movement*; Winston, *Claiming the Real*.
9. Swinson, *Writing for Television*.
10. On *Special Enquiry*, see Swallow, *Factual Television*, 72–77; and Corner, "Documentary Voices," 43–49.

11. Rotha remained at the BBC for only two years; on his unhappy time there, see Boon, *Films of Fact*.
12. Sexton, "Televerité Hits Britain," 431.
13. *Night in the City* is available as part the BFI's DVD release *Visions of Change, Volume 1: BBC, 1951-1967* (2015).
14. Broadcast 7.30–8.00pm, 1 February 1955.
15. The radio feature transcription comes from Franklin, "Denis Mitchell," which also includes a thoughtful analysis of Mitchell's work as a radio producer.
16. Quoted in Miall, "Obituary: Denis Mitchell."
17. Purser, "Think-tape," 109. *People Talking* was an occasional radio series that ran from May 1954-December 1958. Both the radio version of *Night in the City* and *The Talking Streets* were produced as episodes for the series.
18. Swallow, *Factual Television*, 195.
19. Bridson, *Prospero and Ariel*, 247.
20. Reisz, "Denis Mitchell's Television Films," 52.
21. Gilliam, "Aspects of the Feature Programme," 101.
22. Dupin, "Free Cinema," no page; booklet published as part of the DVD set *Free Cinema*, London: BFI, 2006. The set includes *Momma Don't Allow*. A comparison of Mitchell and Donnellan's films with Free Cinema documentaries indicates that more attention could productively be paid to the intermedial relationships between radio, television and the cinema in the post-war years.
23. *Morning in the Streets* is available as part the BFI's DVD release *Visions of Change, Volume 1: BBC, 1951-1967* (2015).
24. Corner, "Documentary Voices," 53.
25. Donnellan, *We Were the BBC*, 25.
26. In his estimation, his work was 'completely lacking in substance or intellectual rigour' (Donnellan, *We Were the BBC*, 28).
27. Ibid., 39.
28. This was after rejecting an attachment with Ealing's film editors; Donnellan, *We Were the BBC*, 41.
29. Donnellan, *We Were the BBC*, 43.
30. Ibid., 46.
31. Ibid., 48.
32. Ibid., 56.
33. Ibid., 36.
34. This was not his first film. That honour goes to *Home Town: Warwick* (1956), in which Donnellan was responsible for shooting 'uncuttable picture-postcard shots' taken with a 'double-camera': a BBC-designed hybrid of camera and magnetic recorder that Donnellan remembers was 'as manoeuvrable as Birmingham Town Hall and not nearly as handsome or useful.' Donnellan, *We Were the BBC*, 40.
35. Donnellan, *We Were the BBC*, 58.
36. Franklin, "Documenting the Margins," 20.
37. Drabble, *Black Country*, 21.
38. Franklin, "Documenting the Margins," 23.
39. Donnellan, *We Were the BBC*, 64.

40. Ibid.
41. Long, "Representing Race and Place," 268.
42. Franklin, "Documenting the Margins," 8.
43. Colin Moffat, writing in *Contrast Television Quarterly*, cited by Franklin, "Documenting the Margins," 8.
44. Particularly affecting is the voice of a West Indian man who would like to 'get up a game of cricket' with his new neighbours but 'doesn't know who to ask' for an invitation.
45. Donnellan, *We Were the BBC*, 242.
46. Reviews were poor; Cox, *Set Into Song*, 260.
47. Donnellan, *We Were the BBC*, 242.
48. Howkins, "History and the Radio Ballads," 92.
49. Donnellan, *We Were the BBC*, 243.
50. Franklin, "Documenting the Margins," 52.
51. Donnellan, *We Were the BBC*, 254–57.
52. Ibid., 301.
53. Ibid., 272.

Television and Radio Programmes Discussed

The Big Hewer, adapted and produced by Philip Donnellan, 22.10, 50 minutes, BBC Two, 9 May 1974.

The Big Hewer, music and direction by Peggy Seeger, songs by Ewan MacColl, and produced by Charles Parker, 20.30, 50 minutes, BBC Home Service, 18 August 1961.

Coventry Kids, produced by Philip Donnellan, 21.15, 40 minutes, BBC Television, 15 November 1960.

Eye to Eye: Night in the City, devised, produced and directed by Denis Mitchell, 20.45, 30 minutes, BBC Television, 14 June 1957.

The Fight Game, adapted and produced by Philip Donnellan, 22.10, 50 minutes, BBC Two, 11 June 1973.

The Fight Game, by by Ewan MacColl, Peggy Seeger and Charles Parker, 20.00, 55 minutes, BBC Home Service Basic, 3 July 1963.

Gone for a Soldier, produced by Philip Donnellan, 20.35, 105 minutes, BBC Two, 9 March 1980.

Home Town: Warwick, written by Philip Donnellan and produced by Barrie Edgar, 20.00, 45 minutes, BBC Television, 10 August 1956.

Joe the Chainsmith, written and produced by Philip Donnellan, 21.30, 30 minutes, BBC Television, 7 November 1958.

Momma Don't Allow, directed by Karel Reisz and Tony Richardson, 16 mm, BFI Experimental Film Fund, 1956.

Morning in the Streets, directed by Denis Mitchell and Roy Harris, 21.25, 35 minutes, BBC Television, 25 March 1959.

Omnibus: Pure Radio, written and directed by Philip Donnellan, 22.15, 50 minutes, BBC1, 3 November 1977.

People Talking: Night in the City, arranged and produced by Denis Mitchell, 19.30, 30 minutes, BBC Home Service North, 1 February 1955.

People Talking: The Talking Streets, arranged and produced by Denis Mitchell, 8.00pm, 30 minutes, BBC Home Service Basic, 27 October 1958.

The Shoals of Herring, adapted and produced by Philip Donnellan, 21.25, 60 minutes, BBC Two, 15 September 1972.

Singing the Fishing, by Ewan MacColl and Charles Parker, BBC Home Service, 19.30, 55 minutes, 16 August 1960.

Bibliography

Aitken, Ian. *Film and Reform: John Grierson and the Documentary Film Movement*. London: Routledge, 1990.

Anthony, Scott, and James G. Mansell, eds. *The Projection of Britain: A History of the GPO Film Unit*. London: BFI/Palgrave Macmillan, 2011.

Boon, Timothy. *Films of Fact: A History of Science in Documentary Films and Television*. London: Wallflower Press, 2008.

Bridson, D. H. *Prospero and Ariel: The Rise and Fall of Radio*. London: Victor Gollancz, 1971.

Briggs, Asa. *The History of Broadcasting in the United Kingdom: Sound and Vision*. Oxford: Oxford University Press, 1979.

Briggs, Asa. *The History of Broadcasting in the United Kingdom: Competition 1955–1974*. Oxford: Oxford University Press, 1995.

Corner, John. "Documentary Voices." In *Popular Television in Britain: Studies in Cultural History*, edited by John Corner, 43–49. London: BFI Publishing, 1991.

Cox, Peter. *Set into Song: Ewan MacColl, Charles Parker, Peggy Seeger and the Radio Ballads*. London: Labatie Books, 2008.

Donnellan, Philip. "We Were the BBC: An Alternative View of a Producer's Responsibility, 1948–1984." Unpublished typescript, dated 1992.

Drabble, Phil. *Black Country*. London: Robert Hale Limited, 1952.

"Feature Programmes". *Radio Times*, August 31, 1928, p. 367.

Franklin, Ieuan. "Denis Mitchell – Seen but Seldom Heard." July 10, 2015. https://ifranklinblog.wordpress.com/2015/07/10/denis-mitchell-seen-but-seldom-heard/ Accessed October 26, 2017.

Franklin, Ieuan. "Documenting the Social and Historical Margins in the Films of Philip Donnellan." *Littératures, Histoires des Idées, Images, Sociétés du Monde Anglophone – Literature, History of Ideas, Images and Societies of the English-speaking World* 12, no. 1 (2014). http://lisa.revues/org/5606. Accessed May 15, 2014.

Franklin, Ieuan. "Think-Tape: The Aesthetics of Montage in the Post-War Television Documentary." In *Television Aesthetics and Style*, edited by Jason Jacobs, and Steven Peacock, 289–307. London: Bloomsbury, 2013.

Gilliam, Laurence. "Aspects of the Feature Programme." *BBC Quarterly* II, no. 2 (Summer 1947): 100–104.

Hendy, David. "Painting with Sound: The Kaleidoscopic World of Lance Sieveking, a British Radio Modernist." *Twentieth Century British History* 24, no. 2 (2013): 169–200.

Howkins, Alun. "History and the Radio Ballads." *Oral History* 28, no. 2, Memory and Place (Autumn 2000): 89–93.

Long, Paul. "Representing Race and Place: Black Midlanders on Television in the 1960s and 1970s." *Midland History* 36, no. 2 (2011): 262–277.

Madsen, Virginia M. "Radio and the Documentary Imagination: Thirty Years of Experiment, Innovation, and Revelation." *The Radio Journal: International Studies in Broadcast and Audio Media* 3, no. 3 (2005): 189–198.

Miall, Leonard. "Obituary: Denis Mitchell." *The Independent*, October 4, 1990.

Pettitt, Lance. "Philip Donnellan, Ireland and Dissident Documentary." *Historical Journal of Film, Radio and Television* 20, no. 3 (2000): 351–365.

Purser, Philip. 1961. "Think-Tape: A Profile of Denis Mitchell." *Contrast* 1 (2): 108–114.

Reisz, Karel. 1959. "On The Outside Looking In: Denis Mitchell's Television Films." In *International Film Annual No. 3*, edited by William Whitebait, 50–54. London: John Calder.

Reynolds, Stanley. "Gone for a Soldier." *The Times*. March 10, 1980.

Scannell, Paddy. 1986. "'The Stuff of Radio': Developments in Radio Features and Documentaries before the War." In *Documentary and Mass Media*, edited by John Corner, 1–28. London: Edward Arnold.

Sexton, Jamie. 2003. "Televerité Hits Britain: Documentary, Drama and the Growth of 16 mm Filmmaking in British Television." *Screen* 44 (4): 429–444.

Sussex, Elizabeth. *The Rise and Fall of British Documentary*. Berkeley, CA: University of California Press, 1985.

Swallow, Norman. *Factual Television*. London: The Focal Press, 1966. pp. 72–77.

Swann, Paul. *The British Documentary Film Movement 1926–1946*. Cambridge: Cambridge University Press, 1989.

Swinson, Arthur. *Writing for Television*. London: A. & C. Black, 1955.

Thomas, Jeanette. "Gilliam, Laurence Duval (1907–1964)," *Oxford Dictionary of National Biography*. Oxford University Press, 2004. http://www.oxforddnb.com.ezproxy2.londonlibrary.co.uk/view/article/40890 Accessed July 6, 2017.

Winston, Brian. *Claiming the Real: The Griersonian Documentary and its Legitimations*. London: BFI, 1980.

AFTERLIVES OF BBC RADIO FEATURES

Amanda Wrigley

BBC Radio feature programmes written by established literary figures in the mid-twentieth century enjoyed richly creative afterlives across many decades and in a variety of media—for example, in print, as commercial recordings, in theatre performance and on television—as well as in a succession of new productions on radio. This activity kept works alive in the public imagination beyond the ephemeral moment of first broadcast and, it is argued, contributed to the sense (for audiences past and scholars present) of an informal canon of literary radio features. This essay explores the intermedial afterlives of three such literary features—Sackville-West's The Rescue *(1943), MacNeice's* The Dark Tower *(1946) and Thomas'* Under Milk Wood *(1954)—in order to demonstrate the significance of the form as a site for exploration of social issues, politics and cultural life. The essay concludes with a call for more wide-ranging attention to the protean feature form, including work that may not have persisted in the schedules, or had rich, intermedial afterlives, but that may still offer significant insights into the history of social, political and cultural life in mid-twentieth-century Britain.*

This essay considers feature programmes written for and first broadcast by BBC Radio networks in the middle decades of the twentieth century which had rich intermedial afterlives following the moment of their first broadcast.[1] The particular focus is on 'literary' radio features—those taking full advantage of the modernist aesthetics of imaginative radio, often written by established poets and playwrights—published soon after first broadcast by houses such as Faber, Allen & Unwin and Secker & Warburg. Repeat broadcasts and new radio productions of features, together with published versions of the scripts, kept these works alive in the minds of listeners beyond the ephemeral moment of first broadcast. When technology and economics permitted the mass circulation of audio formats, sound recordings from the archives could be heard outside of the radio schedule.[2] In addition, some examples of this most radiogenic of forms have also been, over the years, adapted and realised in other kinds of performance—including staged readings, theatre performances, television productions, films, operas and orchestral concerts. Archival recordings of literary radio features have recently reached new audiences via public listening events held at the British Library where enthusiastic listeners demonstrate sustained interest in examples from this body of work.

These afterlives—re-engagements with radiogenic work across a variety of media over a number of decades—have doubtless helped to construct the sense that there is a central corpus or canon of literary radio features emanating from creative writers and producers working in the mid-century BBC. For those who know something about the output of the BBC Features Department works by established literary figures (such as *The Dark Tower* and *Under Milk Wood*) are likely to be amongst those that instantly spring to mind. Scholarly attention has, naturally, been readily attracted to the output of high-profile literary figures, especially when their radio work has been revived and re-performed regularly on radio and has undergone fascinating intermedial metamorphoses (whether in print or other performance forms). Biographers, critics and literary scholars have increasingly come to appreciate that radio features are not an inconsequential dimension of writers' work; rather, they can offer vital clues towards a full understanding of stylistic development.[3] Historians, too, are interested in the bigger literary names not only for the quality of their radio-writing but also for the weight their radio output adds to the general case for radio as a medium worthy of serious study. Much important work has been done, and the valuable links forged between literary scholars and those with a more media historical perspective are mutually enriching.

The fact remains, however, that a good deal of the output of the Features Department remains under-explored, whether as a result of long-term invisibility of works owing to absence of a creative afterlife, the lack of fame attached to the writer's name or issues relating to the preservation and accessibility of archival material. Even Louis MacNeice—whose radio writing has been published, investigated, interpreted and appreciated perhaps as much as, if not more than, any writer of features—was so prolific that scores of his features remain largely unknown.[4]

The output of certain radio-writers has therefore achieved a kind of canonical status as a result of their literary standing beyond radio and because of the (not unrelated) revival and intermedial recirculation of their writing for radio. This essay seeks to illustrate the ways in which the sense of an informal canon of literary radio features may have arisen —for audiences past and scholars present—through an exploration of the rich intermedial afterlives of three literary features: Edward Sackville-West's *The Rescue* (1943), Louis MacNeice's *The Dark Tower* (1946) and Dylan Thomas' *Under Milk Wood* (1954). First, I outline aspects of the symbiotic relationship between radio and print forms—from the *Radio Times*, to *The Listener*, BBC pamphlets, books and libraries. Radio's close relationship with print—as demonstrated in all three examples—suggests that in some sense radio was ephemeral only in theory. Not only did the chronological framing of the audio experience (at least as it was planned up to the moment of printing) leave its trace in published listings (a 'modernist text' in itself, argues Kate Lacey in this special issue) and the bare transcript of some features, sometimes with descriptions of sound, exist in magazines or books, but radio programmes were also remembered. They lingered in the imagination, became part of the collective fabric of cultural experience and stimulated the widely held desire for repeated engagements with interesting content. The *experience* of radio, then, is (or, at least, can be) quite the opposite of ephemeral. It was partly in response to this long-lasting imaginative impact of radio and the appetite it created for repeated or recalled experience that many features made it into the permanence of print. Writers with established relationships with publishers were, it seems, easily able to satisfy their own keen wish to see their radio-writing printed in books and the listener-reader's appetite for features-in-print was strong.

The question of how we may achieve a fuller knowledge of features in this period, one that extends beyond the relatively well-known and accessible works, lies at the heart of this study. The features that have gained a certain profile over the decades, often via re-circulation in different performance forms and media, including print, demonstrate the significance of the form as a site of exploration of contemporary social issues, politics and cultural life. More wide-ranging attention to this protean form, beyond the informal features canon that has accumulated over the decades, may yet reveal much more of significance for the social and cultural life of Britain in the middle decades of the twentieth century.

Print Radio

> [...] printed book and radio transcript popularize each other; it is sometimes a two-way traffic.[5]

The relationship between BBC Radio and print was, from the start, symbiotic. The daily schedules published in the weekly magazine *Radio Times*, for example, were fundamentally important in communicating to the reader, a potential listener, what the BBC planned to broadcast and when. Programme listings were often accompanied by some words or sentences of description and they were sometimes illustrated by photographs or evocative line-drawings to attract the eye and fire the imagination.[6] The pages preceding the schedules were also used as a space for, *inter alia*, long-form articles about upcoming programmes: indeed, the first page-length article in the first number of the *Radio Times* in 1923 was a transcript of a 'recent talk broadcast'.[7] *Radio Times* was an incredibly successful publication—indeed, it kept the BBC financially afloat—but its success meant that, as Asa Briggs details, 'the Corporation was involved in frequent quarrels with the press' which viewed broadcasting as a rival.[8]

From 1929, the BBC periodical *The Listener* offered a new space for printing talks of significance (such as the National Lectures, which began in 1929 with Poet Laureate Robert Bridges)—indeed, its 'primary function [was] to satisfy the constant demands made for the text of broadcast talks after delivery'.[9] This report of 'constant demands' is evidence for the proportion of the listenership who keenly desired a more permanent record of what they had heard, and the most readily available such 'record' available for circulation at this time was the printed word. Opposition to the establishment of this new journal was strong in the world of publishing, with the leading newspaper associations and societies petitioning even the Prime Minister to intervene to stop it from reaching the printers. The settlement to which the BBC committed stipulated that *The Listener* take minimal advertising revenue and publish mostly print versions of broadcast programmes; although this did not in practice limit its ambitions and achievements as a thinking person's journal.[10]

As was the case also for the National Lectures, pamphlets were also regularly printed for general talks, libretti and concert 'programmes' for classical music broadcasts. This was akin to the way that the large amount of radio programming for schools from the 1920s regularly relied on the publication, sale and circulation of Aids to Study pamphlets in order to provide essential background, context and further resources for both teachers and schoolchildren to get the most out of broadcasts.

The BBC also published its own books (including its annual 'yearbook') and cultivated strong working relationships with external publishing houses—notably with Penguin. Documentary evidence in the BBC Written Archives Centre and the Penguin Archive suggests that the two often operated hand-in-hand: words heard on the air often were about to be, or had recently been, printed on the Penguin page.[11] It was common practice for Penguin to send in new titles for distribution to BBC staff; and, over two decades, Allen Lane, the founder of Penguin Books, contributed to a variety of radio programmes (for example, a careers programme for schools on the publishing industry in 1947 and a programme on the 21st birthday of Penguin Books in 1956).[12] In 1969, *The Times Literary Supplement* considered Penguin to be 'the nearest thing to a BBC-type publishing house impelled by an educative zeal',[13] a reference to ideological similarities between Allen Lane and John Reith, Director-General of the BBC to 1938.

In more general ways, too, did radio and print reinforce each other's cultural and (self-)educational offerings. Asa Briggs considers that from the 1920s 'more than one local librarian in his annual report referred to wireless as a "new ally [...] creating and deepening the interest of the public in the higher forms of literature"'.[14] Mark Pegg notes that, although the characteristic informality of the educational effect of broadcasting means that it is difficult to measure its impact with precision, 'radio could provide the incentive to read a book cited in a broadcast: there was evidence that novels dramatised for the radio were quickly put under heavy demand' in libraries. Furthermore, public librarians supported the activities of wireless listening and discussion groups with related book displays and direct sales of BBC publications.[15]

Features on the Page

Several essays in this special issue of *Media History* highlight this mutually supportive relationship between radio and print. Indeed, print afterlives of radio feature programmes are a distinctive strand of discussion: for example, Alex Lawrie notes how the 1946 feature programme *A Room of One's Own* sought to demystify Virginia Woolf for the Home Service audience, encouraging listeners to read the novelist's work—an instance which exemplifies the promotion of reading as an inadvertent consequence of and therefore inseparable from the BBC's broadly educative aims. Lawrie also documents the adaptation of a famed work in print—David Jones' epic poem of 1937, *In Parenthesis*—into an acclaimed feature, with sound effects, music, drama, as well as verse. Aasiya Lodhi recounts how Louis MacNeice in the 1950s approached Faber (which had already published his poetry and features such as *Christopher Columbus* and *The Dark Tower*) with an idea for a 'not too usual kind of travel book' from a selection of the features he had written and produced on overseas topics.[16] Alex Goody notes how the BBC's arrangement with the Bodley Head in 1940 to publish six episodes of the features series *The Shadow of the Swastika* (1939-40) pre-empted efforts by staff writers A. L. Lloyd and Igor Vinogradoff to serialise the works in a newspaper or to secure a book contract on their own terms. Goody also writes of how Terence Horsley drew on material forthcoming in his book *Norway Invaded*, co-authored with James Tevnan, for his radio feature *Narvik* (1940) and how Cecil McGivern's 1941 feature *The Battle of Britain* was published later that year by Allen & Unwin alongside five of his other wartime pieces in the well-illustrated *Bomb Doors Open and Other Radio War Features*.

In 1950, Laurence Gilliam, the longstanding head of the Features Department, himself edited a book of scripts from fifteen 'outstanding' feature programmes. Published by Evans Brothers Limited, by arrangement with the BBC, it was intended for 'those who wish to re-live the enjoyment of a favourite feature and [...] those who were unfortunate enough to miss the programme'.[17] Gilliam's introduction to this volume makes clear that the primary purpose of the collection is to provide 'a more permanent record' of ephemeral radio writing of high quality. He says more about the process of adaptation of sound to print undertaken in his volume:

> It may at first seem perverse to attempt to transform material originally created to be heard into material to be read. The test is in the reading. There seemed little point in reprinting these programmes in their original radio scenario form, as prepared for the actors and producer. All the pieces printed here have been rewritten by their original authors at considerably shorter length than the original broadcast. It is hoped that the essence of each has been retained.[18]

Related to the modifications made from script-as-broadcast to feature-as-printed, some items—such as *The Man from Belsen* (1946) by Leonard Cottrell and *British Justice* (1947) by Jennifer Wayne—are preceded by an authorial note on the shape, structure and style of the radio broadcast of their work, indicating a desire to offer a stronger sense of the audio experience.[19]

In MacNeice's introductions to his books of radio scripts, several of which were published by Faber, he gives a range of reasons for seeking publication: amongst them is the (clearly common) desire for a more permanent record of his work for the medium, since 'like all radio writers I feel frustrated each time a script has been broadcast', 'because it is "over"'.[20] He believes that those he has published are, quite simply, 'worth reading'; but, also, he uses these introductory essays in large part to 'throw some light on the medium', speaking, it appears, to writers who may themselves consider writing for radio.[21] He notes how the publication of features 'may interest some members of the more literary public in a popular art-form which still is an art-form', alluding, perhaps, to the disdain in which some of the intelligentsia famously held mass media.[22]

The introduction to the 1944 Faber edition of *Christopher Columbus* is important for MacNeice's nascent thoughts on radio as a medium for creative expression for writers. He emphasises the crucial distinction between writing for readers and writing for radio: for example, 'those subtleties which the ordinary writer uses in rhythm or phrasing—or thought—will often be superfluous and sometimes detrimental to the radio writer'.[23] In encouraging writers 'to forget about "literature" and to concentrate upon sound', invoking the model of the Homeric or Icelandic bard, MacNeice underlines the importance of sound effects and, crucially, music.[24] 'What music can very often do', he considers, 'is to compensate for the lack of the visual element, to establish an emotional atmosphere or to register a change of mood more vividly, and more quickly, than words alone'.[25] Whereas Gilliam's volume cut down and in other ways adapted its radio works for the page, in his 1947 *The Dark Tower and Other Radio Scripts*, for example, MacNeice reprinted 'dialogue [...] as broadcast but the directions have been re-written and considerably amplified', to give as much of a sense as possible of the aural experience.[26] Similarly, in his 1945 Secker & Warburg edition of *The Rescue* Edward Sackville-West 'left the script just as it stood, music cues and all, instead of attempting to refurbish it in the interests of what might seem superior readability', citing 'its suggestive value for other writers' as a motivation here.[27]

Although the vitally important dimension of sound was impossible to reproduce in print, books could (alongside 'directions' and 'music cues') offer additional resources and attractions—paratexts which served to inform, illustrate and contextualise. In addition to the kind of thought-provoking methodological introductions that writers including MacNeice and Sackville-West offered their readers, volumes of scripts published in the mid-twentieth century also offered notes (on e.g. historical background and sources), illustrations (whether visual 'actuality' in the form of historical photographs as in McGivern's *Bomb Doors Open* and Gilliam's *BBC Features*, or artistic work such as in *The Rescue* which, as discussed below, printed six full-page drawings commissioned from Henry Moore). This additional material was far more than could be published in the *Radio Times* and *The Listener* before or after broadcasts. Book publication thus gave readers who had heard a feature a stimulating reminder of that experience, together with an enriched context that could be enjoyed even by those who had not been able to 'listen in' to that audio event.[28] In the 1940s, especially, it is clear that adapted and enriched features scripts were published by those with a degree of literary standing also, in part, as a reflection on the state-of-the-art and as an address to writers who were yet to be persuaded of the potential and power of the form.

'*Very close to mike at first, then gradually emerging from the frame*': *The Rescue's* Print and Musical Afterlives

Novelist and music critic Edward Sackville-West (1901–1965) wrote *The Rescue* as a radio feature programme whilst on the staff of the BBC in the Second World War. It was broadcast in two parts on 25 and 26 November 1943. The idea for this dramatic feature programme had come from the Director of Music Arthur Bliss who planned to write a radio opera based on Homer's *Odyssey*. *The Rescue* was the first substantial creative treatment of Homeric epic on BBC Radio—a truly radiogenic piece. It is modelled on the telling of Odysseus' eventual arrival home on Ithaca after the Trojan War, only to find, and slay, the suitors besieging his wife Penelope, but the characters are developed substantially in line with Sackville-West's desire 'to haul them back from the remote limbo created by the unreal vocabulary of academic translation'.[29] Music was foundational to the conception of the piece and Benjamin Britten (1913–1976) was commissioned to provide a substantial score.[30] In a half-page *Radio Times* article, Sackville-West described this radio work as 'not simply a play with incidental music stuffed into the joints or used to add colour here and there, but an attempt to unite speech, song and symphonic music in a form which only the microphone can manage'.[31] Indeed, the collaboration of Sackville-West and Britten made a distinctive exploration of the dramatic potential of radio features.

In his *Radio Times* article, illustrated by a poignant drawing of Odysseus asleep in a cave by famed artist Eric Fraser, Sackville-West offers a synopsis of the play, discusses the reason for the focus on this episode in the epic poem and his treatment of the characters, and encourages the reader, a potential listener, to link up the themes of *The Rescue* with wartime geopolitics. Within two years the text had been published by Secker & Warburg, prefaced by a ten-page 'Preamble' in which Sackville-West discusses contemporary resonances in greater depth: 'For the startling parallel between the story, as I have told it, and the present state of Greece, I make no apology. It was too obvious to require underlining'.[32] As noted above, Sackville-West 'left the script just as it stood, music cues and all' which are evocative and revealing of technique: for example, *'Whip cracks and screams of women'*, *'Distant sound of four Naiads singing'* and *'Very close to mike at first, then gradually emerging from the frame'*.[33]

The volume is illustrated by six full-page drawings especially made by Henry Moore in 1944, four of which were printed in colour and two in black and white. Announcing the 1945 publication in *Radio Times*, a critic notes that 'the publishers explain that it was outside the scope of this book to print the score, but by way of compensation they offer reproductions of six designs by Henry Moore—the first of his that have appeared as illustrations to a text'.[34] That Moore should have provided these compensatory illustrations is unexpected: he was, of course, primarily a sculptor and classical imagery was for him 'never a dream of a lost idea that might be recovered if only we knew how' but an 'existential classicism, toughened and reduced'.[35] His extraordinary drawings, together with others from his sketchbook for *The Rescue*, are an important wartime intertext for his Shelter Drawings and his reclining nudes and they continue to resonate, being featured in recent exhibitions such as 'Moore and Mythology' (Henry Moore Foundation, Perry Green, 2007) and 'The Mythic Method: Classicism in British Art' (Pallant House Gallery, 2016).

The 1943 première of *The Rescue* was followed by at least six new productions on BBC Radio (in 1948, 1951, 1956, 1962, 1973 and 1988). Val Gielgud's 1956 production was one of several feature productions to be showcased in a two-week Festival of Radio Drama broadcast on the Home Service in May. This festival presented 'plays conceived and written for broadcasting over the past thirty years and widely recognised as classic examples of the art of radio drama'.[36] The fluidity in the terminology here—'radio drama', 'features', 'plays'—was ever thus (see the introductory essay in this special issue), but it is notable that these words, lauding radiogenic features as *art*, are written by the culturally traditionalist Val Gielgud (head of radio drama and features, later just drama), a staunch advocate of theatre play production on radio. This kind of 'festival' in itself did much to create and maintain the sense of a canon of literary features over the decades, and the selection in 1956 included the usual suspects, such as *The March of the '45* by the influential producer D. G. Bridson and Louis MacNeice's *The Dark Tower*.

Fresh attention was paid to Britten's score from 1988, following the radio production by Chris De Souza and Ian Cotterell which featured a five-minute 'interval' reading from Michael De-la-Noy's new autobiography of Sackville-West;[37] it was immediately followed by Britten's 'Serenade' (Opus 31), composed in 1943 and dedicated to Sackville-West. Stimulated by his work on this production, De Souza went on to compile a 45-minute concert version of the music for narrator, soprano, mezzo-soprano, tenor and baritone soloists with orchestra, which retained Athene as narrator. This new work, 'The Rescue of Penelope', premièred at Snape Maltings in Aldeburgh on 23 October 1993, fifty years after the first radio broadcast of *The Rescue*; and Warner Elatus issued a recording of it in 1995.[38]

' ... voices dwindle away and a tolling bell grows up out of the distance': *The Dark Tower* in and out of Print

Britten also composed the music for MacNeice's *The Dark Tower* which was written in the spring and summer of 1945 and first broadcast on the BBC Home Service on 21 January

1946. MacNeice dedicated the 1947 Faber edition of the play to Britten and concluded his introduction with the observation that 'Without his music *The Dark Tower* lacks a dimension'.[39] The protagonist is the young Roland, a reluctant hero sent off by his mother on a quest which he does not comprehend and does not take up readily, his father and brothers having before him set out on the same quest never to return.[40] This play, written and produced near the end of the Second World War, offers much food for thought on notions of authority, duty, and free will whilst also attempting to do 'justice to the world's complexity': 'I have my beliefs and they permeate *The Dark Tower*', wrote MacNeice, 'But do not ask me what Ism it illustrates or what Solution it offers. You do not normally ask for such things in the single-plane work; why should they be forced upon something much more complex?'[41] It was a work which immediately appealed across Europe, being translated into French for broadcast in Belgium and France; and, together with his feature *The Golden Ass*, it was translated into Italian for a production which benefited from MacNeice being present at rehearsals.[42]

As noted in the introduction, above, Louis MacNeice's writing for radio (much of which he produced himself) has received a great deal of scholarly attention. His features are amongst the best known examples of the form, with many of them available in print —and yet, because he was so prolific, a great number of the at least 160 scripts he wrote for broadcast remain largely unknown.[43] Like many literary figures, he was keen on publishing his features and took advantage of established relationships with editors. Soon after starting at the BBC in 1941, he offered Faber (which published his poetry) examples from his radio work. In 1942, writing to T. S. Eliot, he mentioned that 'merely on the page' his recently broadcast feature *Christopher Columbus* 'would read intelligibly and effectively'; Eliot agreed, and it was published in 1944.[44] In 1945, he offered more scripts, including *The Dark Tower*: 'I should of course want to re-write the directions, supplying at least an illusion of the visual element. [...] If you fancy publishing these five in a volume, I can straight away fall to and make a readable version of them'.[45] Accordingly, these were published in 1947. Some fifteen years later, he wrote to Charles Monteith at Faber to ask:

> I suppose you would not be interested in any new radio plays? [...] I was always sorry that, while you have kept both *Christopher Columbus* and (God help us) *Out of the Picture* in print, you let *The Dark Tower*, which I consider much more interesting than either, go out of print. Schools and amateur theatrical societies etc. still keep asking me for permission to give readings or performances of it and I imagine the same would happen with these more recent pieces.[46]

As well as illustrating the easy relationship here between MacNeice and colleagues at Faber, this letter is evidence for the way that *The Dark Tower* was enjoyed by audiences not just as something heard in successive productions on radio (in 1946, and, for example, in 1950, 1956, 1963, 1966, 1973, 1995 and later), kept in mind and read in book form, but also as something that they may themselves perform. The continuing place of *The Dark Tower* in the public imagination is illustrated by the 2017 'recreation' of the original broadcast in front of a live audience under the tower of Orford's medieval church, with Britten's music played by the BBC Concert Orchestra, a recording of which was later broadcast on BBC Radio 3.[47]

Further scripts were published in the years following his death in 1963. *The Dark Tower* was, as was his wish, reprinted in 1964; and in the same year Faber published a volume comprising *The Mad Islands* (broadcast in 1962) and *The Administrator* (1961). In 1969 the BBC published *Persons from Porlock and Other Plays for Radio*. W. H. Auden, who selected the scripts and wrote a Foreword for this volume, based his selection on the criterion that they 'were so well suited to radio that I cannot imagine them being as successful in any other medium' and, to this end, he chose to omit scripts which reworked existing literature by other authors.[48] In 1993, Alan Heuser and Peter McDonald edited a collection of eight of MacNeice' s plays for the Clarendon Press at Oxford: the stated purpose of this volume was to make accessible the texts of some of MacNeice' s best dramatic works, for whatever medium; included were some already published radio works (such as *The Dark Tower*) as well as some previously unpublished ones (such as *Prisoner's Progress*, a parable play concerning the Second World War prisoner-of-war camps which was broadcast in 1954 and won second place in the Prix Italia competition of that year).

To this point in the publication history of MacNeice' s radio *oeuvre*, then, the focus was on printing what were perceived to be the best works written for radio by one of the acknowledged masters of the form. In 2013, the latest contribution to his published work took a different approach, exploring the various ways he re-worked one particular and recurrent source of material—Greek and Roman history and literature—for radio broadcast over the span of his long career at the BBC.[49] All but one of the eleven scripts in that volume, which I co-edited with classicist Stephen Harrison, had not been printed before: our intention was to expand the known territory and thus get a fuller sense of MacNeice's work in one particular area. Many more of MacNeice's features await similar exploratory attention.

Under Milk Wood: 'The impression remains'

The Welsh poet, writer and broadcaster Dylan Thomas (1914–1953) conceived of *Under Milk Wood* as a piece for performance on radio. Subtitled *A Play for Voices*, it is a linguistically rich documentary fiction which describes the secrets, longings and regrets of many inhabitants of a Welsh seaside town over the course of a single day through a rich tapestry of narration, dialogue, soliloquy and song. Its *joie de vivre* and playful mastery of the form are bittersweet in light of Thomas' untimely death shortly before its first broadcast. The BBC Radio production won the prestigious Prix Italia in 1954 as well as a rapturous reception in the press and amongst the majority of listeners who responded to the BBC's audience research survey. The 94-minute radio première was repeated several times: the Home Service broadcast a shortened version later that year; sixty-minute versions were made for overseas services; and it was translated into at least eight languages for international production. New productions commissioned for BBC Radio followed and it remains a popular work to the present day.

It is in terms of radio that *Under Milk Wood* is primarily considered to be a masterpiece, receiving much critical and academic discussion, but its popular renown derives also from engagements with the work in a range of media following its radio première and its dissemination both in audio formats and in print. Before its first radio broadcast, it was, in fact, performed as a solo reading by Thomas himself in Cambridge, Massachusetts in May 1953, followed by two further readings in New York. After his death six months later further readings and performances were held at the Globe Theatre, the Old Vic, the Festival Hall and the Royal Academy of Dramatic Art.

Douglas Cleverdon, who produced the inaugural radio broadcast in 1954, went on to direct it for the stage with Edward Burnham at the Edinburgh Festival, in the West End and on Broadway across 1956–57. It was at this time common for the BBC to transmit extracts from current theatre productions and for these broadcasts to lead to full television productions. The televising of an extract from one of the Edinburgh performances in 1956 was therefore followed in 1957 by a full BBC television production which, the theatre manager Henry Sherek notes, broke new ground by featuring 'the entire London cast almost immediately after the London closure'.[50] In the years that followed, a second television production was broadcast in the BBC's *Festival* series (1964); Thomas' play also inspired the Stan Tracey Quartet's 1965 *Jazz Suite* and, amongst many other cultural engagements internationally, the play was adapted as film (dir. Andrew Sinclair, 1972) and an opera (composed by Walter Steffens, 1973).

Soon after its first broadcast, Decca issued vinyl recordings of the first production under the Argo label regularly from 1954 and recordings of Dylan Thomas' stage-readings in New York in 1953 were released commercially by Caedmon Records in 1956. In 2001 the BBC issued a BBC Radio Collection CD of the 1963 BBC production, including extracts from the first production from 1954. In 2003, BBC Audiobooks released a two-CD set of 'unique extracts from the BBC archives'. Opening with an introduction from Thomas' biographer Paul Ferris, the collection includes readings of Thomas' poetry and prose, as well as an extract from the 1954 radio production of *Under Milk Wood*, its forerunner *Quite Early One Morning* and *Return Journey to Swansea*. Most recently, in 2014, Pip Broughton directed a star-studded BBC Wales television production for the centenary celebrations of Thomas' birth and the BBC produced *The Road to Milk Wood*, an interactive multimedia digital book on Thomas' life and this work in particular, illustrated with an abundance of archival footage and practitioner interviews. Across the decades, then, opportunities to *hear* Dylan Thomas' *Under Milk Wood* have been, and continue to be, plentiful.

The play could also easily be read. In March 1954, shortly after Thomas' death, J. M. Dent & Sons Ltd published *Under Milk Wood*, an edition of the text compiled by his literary trustee Dr Daniel Jones. New York's New Directions published the American edition the following month. Incomplete and abridged versions had already appeared in the quarterly *Botteghe Oscure* (Rome, 1952), in the February 1954 issue of *Mademoiselle* (New York) and in *The Observer* (7 and 14 February 1954). In the same year, Dent also put into print a volume of twenty-two short stories, sketches and essays from his radio work. Thomas had the idea for this book before leaving for what was to be his last journey to New York; Aneirin Talfan Davies of the BBC Welsh Service edited the collection into shape. 'It is impossible to close the book without regret, without an infinite sense of loss', writes a reviewer in *The Times Literary Supplement*: 'Dylan Thomas as a broadcaster was unique. His place in sound radio was equivalent to Chaplin's place in the silent film'.[51] In 1958, Dent produced an acting edition, including stage directions and lighting cues from the stage production, and, in 1969, *The Growth of Milk Wood*, Douglas Cleverdon's account of the eleven significantly different versions of *Under Milk Wood*, drawing on his close association with both radio and stage productions: 'the process of construction,' he states, 'was hampered, particularly in the later stages, by compromises that were designed to adapt the work for such differing purposes as radio feature, public reading, and stage production'.[52] Cleverdon's purpose was not to work towards a definitive text of *Under Milk Wood*, but to 'analyse the textual variations and to trace the complicated—and occasionally hilarious—development of the script'.[53] Further works on Dylan Thomas abound.

Thomas' untimely death—just as *Under Milk Wood* was coming to life—doubtless gave *Under Milk Wood*'s afterlives added vigour. Its various iterations, through aural performance, commercial recordings, readings, embodied performance, televised productions and film, and the printing of the play in various stages of its evolution in periodicals, newspapers and books, ensured long-lasting renown for this brilliantly impressionistic piece first conceived of as a radio feature. It also serves as a fitting living memorial to the poet's radiogenic expertise and, not inconsequentially, a source of income for the poet's estate. The fact that theatre and television practitioners immediately sought to perform *Under Milk Wood* as live theatre and pre-recorded television says much about how readily its entertainment potential was perceived, and how much audiences enjoyed the work. In an earlier study, I concluded that the language contemporary critics used in reviews of *Under Milk Wood* in its various performative incarnations relied heavily on the discourse of fidelity (in which aesthetic value is perceived in inescapably close relation to the distinctiveness of the communicative strategies of the first, originating medium—in this case, radio).[54] The BBC's research report on *Under Milk Wood*'s television audience (at least that proportion of which that was not distracted by perceived immorality or the absence of plot) held a relaxed position, taking the work at face value and judging it in terms of the production's aesthetics and entertainment value. As one critic who deviated from customary concerns over fidelity put it, *Under Milk Wood* can 'take decoration and expansion without harm. The impression remains'.[55]

Conclusion

Looking back, in 1982, to radiogenic writing for the mid-twentieth-century BBC, Ian Rodger states that 'Radio bred [...] a generation of writers who knew that they must listen while they wrote',[56] echoing MacNeice's encouragement of radio writers 'to forget about "literature" and to concentrate upon sound'.[57] In his introduction to *Christopher Columbus*, MacNeice argued that music could 'compensate for the lack of the visual element'.[58] Yet Rodger's proposition is that:

> [...] any study of the scripts without being able to hear the recordings must necessarily lead to imperfect understanding. Though the meticulously annotated scripts of writers like MacNeice may appear to describe everything that happened in the studio, they cannot convey the magic which was present when the words, the music and the effects were woven together by the actors, the musicians and the sound technicians.[59]

It is obviously true that that printed versions of the scripts of features offer a different experience than listening to the orchestration of the different elements of the sound broadcast. But to speak of 'compensation' and 'imperfect understanding' is to measure the worth of printed versions negatively and to lose sight of the advantages. This is not unlike how early radio practitioners and critics continually defined radio in terms of its 'blindness'.

Scripts, transcriptions, adaptations for the page (or however we may describe printed forms of feature programmes) served different constituencies and different ends.[60] Print enabled writers to achieve a permanent record of and reach wider audiences for their radio work and, in the years of experiment in features' formal aesthetics, radio writers considered it a service to the profession to make their good work widely accessible for their peers. Print also offered opportunities for the enrichment of the aural experience, with historical notes, introductory and interpretative essays, photographic images and illustrative artwork. Clearly, there was a market for them, and a financial benefit to author and publisher. Book-buyers and book-borrowers provided the bulk of the readership and, as we heard from MacNeice, listener-readers enjoyed reading the works, remembering the sound of them and also, for example, *enacting* them in their communities.

Offering both less and also more, the printed form of radio features were valuable experiences in themselves and on their own terms. They were also part of a rich cultural back-and-forth, not only between radio and publishing, but commercially available recordings, films, television productions, and so on. Features have had a surprisingly wide reach via their intermedial afterlives, especially considering the form's much-vaunted medium specificity. The more literary and dramatic work, such as *Under Milk Wood*, has arguably had longer, more complex and textured afterlives than, say, *Bomb Doors Open*. Their authorial pedigree lends them easily to canonisation. When programming the first, experimental series of public listening and engagement events at the British Library, our inclination to select works with which the potential audience might find some kind of 'hook', via the writer's name, for example, or the work's afterlife, was strong. The first series therefore consisted of *The Rescue, The Dark Tower* and *Under Milk Wood*, alongside the radio adaptation of *In Parenthesis* and *The March of the '45*.[61] The *Radio Times* in 1962 declared that 'only *The Dark Tower* [...] can compare with *The Rescue* as an original radio piece in literary quality, sensitively imaginative treatment, and brilliant exploitation of the possibilities of the medium': this kind of language, in seeking to pull in the audience, cannot help but promote the sense that, within features, there was a cultural hierarchy in which some work was exceptional.[62]

These masterpieces, as they are so often termed, of the literary radio form are also extremely valuable examples of the significance of the form as a site of exploration of contemporary social issues, politics and cultural life in the mid-twentieth century. In 1969, a few years after the closure of the Features Department, producer Douglas Cleverdon reminded his readers that 'only a comparatively small proportion of the works produced by the BBC Features Department aimed at the highest standards of poetic achievement. Many of them were actualities, or topical documentaries, or critical features like the series of *New Judgements*'.[63] It feels like the time is approaching when we ought to be widening the lens on the feature form—as this volume strives to do in some respects. The existence of BBC Genome (the project which has put online the programme listings information for BBC radio and television from 1923 to 2009) makes preliminary mapping for such research projects more possible. The British Library's Sound Archive—the public point of on-request access to BBC recordings—is energetically seeking to preserve, especially through digitisation, archival recordings as part of its Save our Sounds programme. And the BBC is striving to make more of its archival riches—its own history—accessible in time for its centenary in 2022. It would be exciting to see, alongside these openings, the emergence of an inter-institutional, interdisciplinary scholarly network dedicated to writing in a collaborative way a fuller history of features to include the large body of work that may not have persisted in the schedules over the decades and may not have had a rich, intermedial afterlife, but which may, as part of a bigger tapestry, reveal significant insights into the social, political and cultural life of Britain in the middle decades of the twentieth century.

Disclosure Statement

No potential conflict of interest was reported by the author.

Notes

1. This essay is not closely associated with Warburgian *Nachleben* theory or subsequent critique (see, e.g., Didi-Huberman, Rehberg, and Belay, "Artistic Survival"); but it is open to the 'messiness' of history in its exploration of the circular movement, metamorphosis and experience of cultural works through time and across media. The related term 'afterlives' is therefore productive; certainly it is preferable to 'adaptation' which remains problematically bound up with language of fidelity (see Johnson, "Adaptation and Fidelity" and Cobb, "Adaptation, Fidelity, and Gendered Discourses").
2. A recent example of the publication of archival sound is the 2006 distribution of the four-CD set *Samuel Beckett: Works for Radio* by the British Library, under licence from the BBC: these include his five pieces created for radio. On the formal categorisation of Beckett's radio works, see essays in Addyman, Feldman, and Tonning, eds, *Samuel Beckett.*
3. Those uninterested in radio as a creative medium often locate radio writing at the periphery of literary figures' *oeuvres*; indeed, critics and obituary writers can be disparaging of writing for the medium. For the range of estimations of MacNeice's radio work in relation to his poetry, see discussion in Wrigley and Harrison, eds, *Louis MacNeice*, 4–5.
4. Wrigley and Harrison, eds, *Louis MacNeice*, sought to fill in some of this picture; much territory remains to be explored.
5. Anon., "The Classics as Best-Sellers."
6. On *Radio Times* artwork see, e.g. Artmonsky, *Art for the Ear*, and Backemeyer, *Eric Fraser*. See Briggs, *The Birth of Broadcasting*, 270 ff. on the way *Radio Times* shaped the BBC's public image.
7. Ratclyffe Dugmore, "Photographing."
8. Briggs, *The Golden Age*, 260ff.
9. Anon., "A 'New Venture'". On printed talks in *The Listener*, see Cohen, "Intermediality", 575; and on the periodical's encouragement of readers to expand their literary horizons, see Lawrie, "'The Appreciative Understanding'".
10. Briggs, *The Golden Age*, 269.
11. On Penguin Classics and radio, see Wrigley, *Greece on Air*, 94–7.
12. Details in BBC WAC, Talks, Allen Lane, file 1, 1939–62.
13. Quoted in Joicey, "A Paperback Guide," 30.
14. Briggs, *The Birth of Broadcasting*, 16, quoting a report written by a Lincoln librarian.

15. Pegg, *Broadcasting*, 209 and 211.
16. Coulton, *Louis MacNeice*, 161–2.
17. Gilliam, ed., *BBC Features*, flyleaf.
18. Gilliam, "Introduction", 13. The permanence of print is described as desirable also by Gielgud, "Preface."
19. *British Justice* was adapted for television presentation in the same year.
20. MacNeice, *The Dark Tower*, 9 and 13.
21. MacNeice, *The Dark Tower*, 9.
22. MacNeice, *Christopher Columbus*, 7. MacNeice admits that before becoming a BBC writer and producer in wartime, he had ('like most of the intelligentsia') been prejudiced against the medium (MacNeice, *The Dark Tower*, 11). See Rodger, *Radio Drama*, 5–8 on the intellectual bias against radio, and radio drama in particular.
23. MacNeice, *Christopher Columbus*, 8.
24. MacNeice, *Christopher Columbus*, 10.
25. MacNeice, *Christopher Columbus*, 18.
26. MacNeice, *The Dark Tower*, 17.
27. Sackville-West, *The Rescue*, 16.
28. On the term 'listen in', see Briggs, *The Birth of Broadcasting*, 222.
29. Sackville-West, "*The Odyssey*."
30. But on the antagonism amongst some BBC staff towards Britten's conscientious objection to active service, which appears to have kept him from being able to conduct his own music for the 1943 production, see Wrigley, *Greece on Air*, 191–2.
31. Sackville-West, "*The Odyssey*."
32. Sackville-West, *The Rescue*, 12–13.
33. Sackville-West, *The Rescue*, 35, 46 and 96.
34. Anon., "Both Sides."
35. Lipsey, *The Spiritual*, 290. For analysis, see Wrigley, *Greece on Air*, 205–6.
36. Gielgud, "A Festival."
37. De-la-Noy, *Eddy*.
38. Porter, "Riches Restored."
39. MacNeice, *The Dark Tower*, 20 (dedication) and 22.
40. MacNeice is reworking Robert Browning's nineteenth-century poem 'Childe Roland to the Dark Tower Came' (the title of which is from *King Lear*, Act 3, scene 4), which focuses on the struggles of an untested medieval knight as he journeys to the Dark Tower where he will die. Browning's poem has inspired many reworkings.
41. MacNeice, *The Dark Tower*, 21 and 22.
42. MacNeice was in Italy in 1947 gathering material for his feature *Portrait of Rome* (memo from George R. Foa to several people including Laurence Gilliam and Louis MacNeice, 1 June 1947, 3, in BBC WAC R19/953).
43. The precise number of scripts he wrote is unknown: see Wrigley and Harrison, eds, *Louis MacNeice*, 1 n. 2 on various estimations.
44. Letter from MacNeice to Eliot, 27 August 1942, in Allison, ed., *Letters*, 445–6 (446).
45. Letter from MacNeice to Eliot, 16 October 1945, in Allison, ed., *Letters*, 461–2 (461, 462).
46. Letter from MacNeice to Monteith, 21 March 1962, in Allison, ed., *Letters*, 690.
47. See https://snapemaltings.co.uk/whats-on/the-dark-tower (last accessed January 10, 2018).

48. MacNeice, *The Mad Islands*; MacNeice, *Persons from Porlock*, 7.
49. Wrigley and Harrison, eds, *Louis MacNeice*.
50. Sherek, "Presenting *Under Milk Wood*." On the stage and television versions, see Wrigley, "Dylan Thomas' *Under Milk Wood*."
51. Watkins, "Dylan Thomas," reviewing Thomas, *Quite Early One Morning*.
52. Cleverdon, *The Growth of Milk Wood*, 1.
53. Cleverdon, *The Growth of Milk Wood*, flyleaf.
54. Wrigley, "Dylan Thomas' *Under Milk Wood*."
55. Granger, "Edinburgh Festival."
56. Rodger, *Radio Drama*, 157.
57. MacNeice, *Christopher Columbus*, 10.
58. MacNeice, *Christopher Columbus*, 18.
59. Rodger, *Radio Drama*, 155–6.
60. Today's scholar is happy to be able to purchase reasonably priced first editions in second-hand bookshops, especially in instances where recordings have been lost or are inaccessible.
61. 'Early Radio Features' (British Library, 2014), co-curated by Amanda Wrigley and Hugh Chignell with the invaluable support of Radio Curator Paul Wilson and in collaboration with In the Dark. My 2016 series, 'Louis MacNeice, Radio Writer and Producer', aimed to bring some of his lesser-known work into the public domain.
62. Gielgud, "*The Rescue*."
63. Cleverdon, *The Growth of Milk Wood*, 8.

ORCID

Amanda Wrigley http://orcid.org/0000-0003-3822-7633

Bibliography

Addyman, D., M. Feldman, and E. Tonning, eds. *Samuel Beckett and BBC Radio: A Reassessment*. New York: Palgrave Macmillan, 2017.

Allison, J. ed. *Letters of Louis MacNeice*. London: Faber, 2010.

Anon. "A 'New Venture'." *The Listener*, January 16, 1929, 14.

Anon. "Both Sides of the Microphone: Homeric 'Melodrama'." *Radio Times*, August 3, 1945, 5.

Anon. "The Classics as Best-Sellers: Influence of Radio on Public Demand." *The Times*, December 29, 1952, 2.

Artmonsky, R. *Art for the Ear: Forty Years of Illustration for BBC Radio Publications*. London: Artmonsky Arts, 2015.

Backemeyer, S. *Eric Fraser: Designer and Illustrator*. London: Lund Humphries, 1998.

Briggs, A. *The History of Broadcasting in the United Kingdom*, Vol. 1: *The Birth of Broadcasting*. Oxford: Oxford University Press, 1961, rev. edn 1995.

Briggs, A. *The History of Broadcasting in the United Kingdom*, Vol. 2: *The Golden Age of Wireless*. Oxford: Oxford University Press, 1965, rev. edn 1995.

Cleverdon, D. *The Growth of Milk Wood*. London: Dent, 1969.

Cobb, S. "Adaptation, Fidelity, and Gendered Discourses." *Adaptation* 4, no. 1 (2011): 28–37.

Cohen, D. R. 2012. "Intermediality and the Problem of the Listener." *Modernism/modernity* 19 (3): 569–592.

Coulton, B. *Louis MacNeice in the BBC*. London: Faber, 1980.

De-la-Noy, M. *Eddy: The Life of Edward Sackville-West*. London: Arcadia, 1988.

Didi-Huberman, G., V. Rehberg, and B. Belay. "Artistic Survival: Panofsky vs. Warburg and the Exorcism of Impure Time." *Common Knowledge* 9, no. 2 (2003): 273–285.

Gielgud, V. "A Festival of Radio Drama." *Radio Times*, May 4, 1956, 3.

Gielgud, V. "Preface." In *Bomb Doors Open and Other Radio War Features*, edited by C. McGivern, 7–10. London: Allen & Unwin, 1941.

Gielgud, V. "*The Rescue*." *Radio Times*, February 22, 1962, 50.

Gilliam, L. "Introduction." In his edited volume *BBC Features*, 9–14. London: BBC, 1950a.

Gilliam, L., ed. *BBC Features*. London: BBC, 1950.

Granger, D. "Edinburgh Festival: *Under Milk Wood*." *The Financial Times*, August 23, 1956, 2.

Johnson, D. T. "Adaptation and Fidelity." In *The Oxford Handbook of Adaptation Studies*, edited by T. Leitch. Oxford: Oxford University Press, 2017.

Joicey, N. "A Paperback Guide to Progress: Penguin Books 1935-*c*. 1951." *Twentieth Century British History* 4, no. 1 (1993): 25–56.

Lawrie, A. "'The Appreciative Understanding of Good Books': *The Listener*, Literary Advice and the 1930s Reader." *Literature & History* 24, no. 2 (2015): 38–52.

Lipsey, R. *The Spiritual in Twentieth-Century Art*. Mineola, NY: Dover, 1988.

Lloyd, A. L., and I. Vinogradoff. *Shadow of the Swastika: A Radio-Drama in Six Parts of the Story of the German National Socialist Party*. London: Bodley Head, 1940.

MacNeice, L. *Christopher Columbus: A Radio Play*. London: Faber, 1944.

MacNeice, L. *The Dark Tower and Other Radio Scripts*. London: Faber, 1947.

MacNeice, L. *The Mad Islands and The Administrator: Two Radio Plays*. London: Faber, 1964.

MacNeice, L. *Persons from Porlock and Other Plays for Radio, with an Introduction by W. H. Auden*. London: BBC, 1969.

McGivern, C. *Bomb Doors Open and Other Radio War Features*. London: Allen & Unwin, 1941.

Pegg, M. *Broadcasting and Society, 1918–1939*. London: Croom Helm, 1983.

Porter, A. "Riches Restored." *The Observer*, April 16, 1995.

Ratclyffe Dugmore, A. "Photographing Wild Animals." *Radio Times*, September 28, 1923, 3 and 5.

Rodger, I. *Radio Drama*. London: Macmillan, 1982.

Sackville-West, E. "*The Odyssey* in Terms of Modern Radio." *Radio Times*, November 19, 1943, 4.

Sackville-West, E. *The Rescue: A Melodrama for Broadcasting Based on Homer's Odyssey*. Orchestral Score by Benjamin Britten. *With Six Illustrations to the Text by Henry Moore*. London: Secker & Warburg, 1945.

Sherek, H. "Presenting *Under Milk Wood*." *Radio Times*, May 3, 1957, 7.

Tevnan, J. and T. Horsley. *Norway Invaded: The First Full Story*. London: Withy Grove Press, 1940.

Thomas, D. *Quite Early One Morning*. London: Dent, 1954a.

Thomas, D. *Under Milk Wood*, edition by D. Jones. London: Dent, 1954b.

Watkins, V. P. "Dylan Thomas and the Spoken Word." *The Times Literary Supplement*, November 19, 1954, 731.

Wrigley, A. "Dylan Thomas' *Under Milk Wood*, 'a Play for Voices' on Radio, Stage and Television'." *Critical Studies in Television: The International Journal of Television Studies* 9, no. 3 (2014): 77–88.

Wrigley, A. *Greece on Air: Engagements with Ancient Greece on BBC Radio, 1920–1960s*. Oxford: Oxford University Press, 2015.
Wrigley, A. and S. Harrison, eds. *Louis MacNeice: The Classical Radio Plays*. Oxford: Oxford University Press, 2013.

AFTERWORD: RADIO MODERNISMS
Features, Cultures and the BBC

David Hendy

The essays contained in this special issue of *Media History* collectively represent an exciting leap forward in our understanding of the intricate ties between radio and modernism.

As Aasiya Lodhi's & Amanda Wrigley's 'Introduction' hints, there was a time in the not-too-distant past when historians of radio and literary scholars gave the impression of pursuing very different concerns. Those trained in the analysis of written texts and knowledgeable about the lives of those who created them were the first to explore in detail just how much the great modernist authors of the early twentieth century—Joyce, Woolf, Pound—were fascinated by the new medium of radio. The focus, however, was very much on how such writers' engagements with broadcasting influenced their *own* craft, or on how these authors were able to capture imaginatively in written form the ripples of change wrought by the arrival of this novel communicative technology, the feel and rhythms of a newly 'mediatised' world. The radio-ness of radio was important, the broadcasting institutions themselves—and the creative practices within them—less so. Unfairly, perhaps, the impression was created that broadcasting was of less interest to literary scholars because it was not seen as a locus of modernist creativity in and of itself. Historians of radio, in the meantime, were traditionally concerned with modernity rather than with modernism: with how radio as a cultural form 'fitted into' a modern society characterised by mass consumption, declining opportunities for face-to-face interactions, expanding leisure activities for those in work, and decisive moves towards universal suffrage. These historians had the literary scholars to thank for providing them with original, highly granular analyses of individual 'texts'—programmes, scripts, and creative works in other media inspired by sound. So they got on with what interested them rather more: mapping the growth and development of radio as a set of institutional structures and practices. Instinctively these scholars saw radio, to varying degrees, as *sui generis*. It spoke of an unfortunate disciplinary tendency towards medium-centricity. But it was a narrowness of attention which did at least mean radio was recognised as its own locus of creativity.

If there was a failure for these two perspectives immediately to connect, it was, perhaps, because so much of what was actually created in radio was—and indeed has continued to be—either routine and banal, or, even when complex and original, has deliberately *effaced* all the effort involved in its making. This is why the work of radio has always appeared to the casual observer to be a rather simple affair, more a case of administration than art. And it is why any radio that has had pretensions to being explicitly modernist has seemed, to many 'mainstream' media scholars, too much of an outlier to deserve sustained attention. In the last resort, it was seen as something grandiose and baroque—a mere *curiosity*, not the *real* business of broadcasting.

Thankfully, the feedback loops in radio's relationship with modernism—and modernism's relationship with radio—have since been well and truly exposed. Literary scholars have revealed their interest in the inner workings of radio; radio scholars have unveiled some long-suppressed literary sensibilities of their own. And the pairing of these two terms, 'radio' and 'modernism' has provided contemporary scholarship with one of the true success stories of that much-sought though rarely achieved thing, cross-disciplinarity. The Introduction to this present collection, again, pays due acknowledgement to the foundational work done in this regard, especially over the past decade. It also reminds us that some earlier work of broadcasting history, published in the 1980s, was perhaps more alert to these cross-currents of influence than we might once have realised. Moreover, it reveals broadcasting itself—and, in particular, the BBC—to be rather less monolithic in its creative processes than was once assumed. The essays here build on—and give added momentum—to this cross-disciplinary trend. They are truly expansive: collectively they stretch our foci of attention in both time and space and introduce us to some entirely new concerns. They also nudge us towards thinking differently about areas we might otherwise have regarded as already fully known. Though they each tackle very different objects of study, taken together they allow the emergence of some common themes—themes that bind even more closely together not just radio and modernism but the two scholarly communities which study them.

For instance, one thing that keeps coming through as a shared point of interest is collage. The simultaneous and enduring fascination among both writers and radio producers with using montage techniques to remove the 'centring' effect of an omnipotent narrator is evident in almost every essay. We see it plainly in Todd Avery's insightful exploration of the metaphorical power of 'waves', both in Tyrone Guthrie's radio work and Virginia Woolf's writings, and again in Alex Goody's thorough exposition of the contrast between the BBC's 'polyphonic' wartime features and Nazi radio's more concentrated styles of vocal address. It is there, too, in Kate Lacey's thoughtful unpicking of the early radio schedule's 'bricolage' character. And we find it in Alexandra Lawrie's sharp analysis of *In Parenthesis* (1946). As always in radio, voices are disembodied. But Lawrie shows how, in the hands of a skilled dramatist, this unavoidable aesthetic reality allowed radio to adopt a constructively playful approach to the hierarchies and relationships of an established text: Douglas Cleverdon's production gave us vocal images juxtaposed in allusive and sometimes jarring ways, allowing for surprising combinations of style and texture—and some exhilarating jumps through history. We hear soldiers and battles from the past as well as the present: spectral armies and the different ages kissing and co-mingling, a splicing together of the earthy voices of the trenches and the mythic language of the Bible and the Welsh bards. This characteristic 'pull-and-push of space and time', which collage so wonderfully enables, also emerges in Aasiya Lodhi's discussion of Louis MacNeice's travel features. The lack of narrative signposting in MacNeice's city portraits allowed the various voices and sounds heard floating out of the radio set to escape the constraining specificities of hard geography, as if they were floating freely across earthly borders. It made for radio that was richly suggestive of the 'uncanny displacements' which seemed to characterise modern life. Here, and in many other examples scattered across these essays, we find in the phenomena analysed a fertile tangle of montage, polyphony, contrapuntalism, disjuncture, multiple time-frames. And lurking behind it all: *subjectivity*. In the twentieth century, thanks to Einstein, radio and literature found themselves in a new and somewhat disorientating age of relativity, where the notion of objective truth and a singular viewpoint was being overthrown in science, in philosophy, in culture. In one sense, when we talk of 'radio modernisms', we are talking about radio and literature —separately and jointly—searching for the right grammar to express this new, complex, and somewhat destabilising reality.

Invoking Einstein has its risks. It suggests something wholly abstract, difficult, rarefied. This would be misleading if we were to neglect another, rather different trope weaving its way through this collection. Yoked closely to the modernist fascination with subjectivity is a shared concern with the *everyday*, and specifically with the kaleidoscopic nature of lived experience. Writers such as Virginia Woolf bent their sentences and paragraphs into a form capable of evoking that rush of sensory stimulation which constituted the moment-by-moment experience of daily life. But radio, always more demotic in its aspirations, and always fully cognisant of its domestic, taken-for-granted character, embraced the 'ordinary' wholesale. Hence even in Alexandra Lawrie's high-art example of *In Parenthesis*, while the structure and form might be complex and demanding, the content is shown to be almost always richly human and multisensory—full of the sights and sounds and fears and hopes and apprehensions of flesh-and-blood characters. For the most part, radio has always represented the ordinary as more than just an averaging out of individual experiences. As Kate Lacey shows, this was manifest from the very beginning in the concept of the schedule. If this constantly updated listing of unending broadcast output was somehow akin to the nation's collective stream of consciousness, it certainly suggested something rather complex, even incoherent. Before routine and ratings tamed it, the schedule embodied a flow of programmes which, differing as they did in subject-matter, tone and difficulty, recognised the multiple identities and dispositions of the many millions of listeners. It is, perhaps, why radio scholars have always been uncomfortable with their object of study being labelled a *mass* medium. Not just radio scholars, either. For as E. Charlotte Stevens and John Wyver show in their essay, some of the techniques of so-called 'pure radio' were to be found in the more artistically-inclined television documentaries of the post-war era. Those made in Britain by Denis Mitchell and Philip Donnellan were, they argue, particularly striking. They shared with radio a delight in sharp, destabilising juxtapositions, which kept a didactic, monologic narrative at bay and replaced it with something more richly allusive and impressionistic; they were also essentially portraits of ordinary, everyday life—works of art profoundly interested in people *as people*, not just people as token embodiments of social processes and institutions. Leonie Thomas' study of Una Marson is pertinent too, in this respect. Her essay reveals an intriguing shift in Marson's own poetry voice as her career progressed, and finally faltered: under the influence of working for BBC radio in London, and what appeared to be a growing consciousness of her 'outsider' status there, her lines became less florid, less enamoured of traditional language, and more reflective of her Caribbean roots: more spiky and hard-edged, more demotic. In radio and literature—again, separately *and* jointly—modernism reveals a consistent effort to give 'voice', imaginatively, to otherwise inexpressible feelings and sensations.

Finally, one is struck by the recurrence throughout these pages of an implicit engagement with rhythm. Though Todd Avery's intention is to draw our attention to the influence of late-Victorian aestheticism in 1920s and 1930s modernism, he does so by revealing the strangely under-reported importance of *musicality*. He invokes afresh that old idea: that all art aspires to the condition of music. And then shows how a range of creative work—notably several of Tyrone Guthrie's plays and both of Lance Sieveking's 'Kaleidoscopes'—stressed their own rhythmic—and sometimes symphonic—qualities. Sieveking's work was subtitled explicitly as 'a Rhythm' in sound, and it was even performed at the BBC's Savoy Hill studios in the manner of a live musical concert, with Sieveking dressing himself in the white tie and tails of an orchestral conductor as he sat hunched over the studio mixing desk. Elsewhere in this collection, we come across the kind of phrases which we might loosely associate with the language of rhythm: we read of pull-and-push, of flow, of counterpoint and lines, of contrast and balance. It is a reminder of the way so many radio practitioners have conceived of their work in terms of 'composing' programmes. And, as Amanda Wrigley shows vividly in her essay on canonical radio features such as *The Rescue* and *The Dark Tower*, this 'fit' between musicality and radio is not just a matter of shared sensibilities or a history of collaboration among the personalities involved (in her particular example, Benjamin Britten, Edward Sackville-West, Louis MacNeice, and Dylan Thomas); nor is it simply an illustration that radio and music are both essentially *time-based* creations. In tracing the 'afterlives' of these features—how they have continued to be performed, re-imagined, and listened-to by new audiences many years after first being broadcast—Wrigley reminds us that radio also has deeper rhythms: rhythms which unfold over the *long durée*. Lance Sieveking once complained of what he called radio's ghastly impermanence. This no longer holds. And it demands of us as scholars that, as far as possible, we situate our 'reading' of modernist radio work in a broader assessment of cultural life, extending far beyond the writing desks of authors and the mixing desks of broadcasters.

To speak of music and of temporal rhythms also surely amounts to a renewed call to pay due attention to the ear: to listen-out for the innate modernisms of radio, rather than reading modernisms into radio. E. Charlotte Stevens and John Wyver refer to the much-used phrase 'pure radio', while Alex Goody and Amanda Wrigley both deploy that other familiar term, the 'radiogenic'. These are telling uses. In equating modernist techniques and styles with radio being its 'true self', we are in effect suggesting, are we not, that while all art aspires to the condition of music, all radio aspires to the condition of modernism? The danger with this position, one might suppose, is not just that it apparently condemns the vast majority of output as not 'really' radio as it should be, but that the definition of 'modernist radio' becomes so all-encompassing that meaning starts to evaporate. For what it is worth, I think there is real virtue in the approach—but only for reasons explored nearly four decades ago, and expressed in words more eloquent than my own, by the novelist Jonathan Raban. He was, he said, suspicious of the term 'pure radio' because it too often meant the creation of something 'technically dazzling' though rather 'arid'. What it *should* mean, he suggested, was something rather broader but seemingly contradictory: radio as an intrinsically 'mongrel form'. One can, for example, hear this mongrel essence in radio's styles of talk:

> In the course of an hour spent as an idle radio listener, twiddling between stations, one drifts from the most elaborate and carefully scripted language through every shade and tone to the most unofficial and unrehearsed grunts and squawks. On radio there is no median register, no particular way of speaking that could be said to represent the medium in neutral gear, ticking over ... Radio is by turns gossipy, authoritative, preachy, natural, artificial, confidential, loudly public, and not infrequently wordless. Its languages bleed into one another. Even the identities of particular channels are constantly being eroded.[1]

'Pure radio', in other words, is radio being its chaotically eclectic self. Which means it is difficult to read Raban's words and not acknowledge that when we talk of radio modernisms we might have to confront the idea that the most profoundly modernist aspect of radio is what we might call—with a nod to Brecht—the medium's 'total effect'. And an exciting—if rather intimidating—conclusion to be drawn is this: that, whether originally inspired by literature or by broadcasting, those of us engaged in studying radio modernisms might need to keep the communication channels open with our 'mainstream' media colleagues, and engage more fully with the mosaic of phone-ins, quizzes, magazines and pop music shows that constitute 'ordinary' output. For if Raban's words still hold true, it could just be there, perversely—beyond the classic texts, and amidst the banalities of cheap, mass-produced, daytime radio—that we find the true modernist soul of this endlessly fascinating medium.

Notes

1. Jonathan Raban, 'Icon or symbol: the writer and the "medium"', in P. Lewis (ed.), *Radio Drama*, London: Longman, 1981, pp. 78–90: 79, 86–87.

Index